D0948612

SMALL BOATS AND DARING MEN

C&C

CAMPAIGNS & COMMANDERS

GREGORY J. W. URWIN, SERIES EDITOR

CAMPAIGNS AND COMMANDERS

Small Boats and Daring Men

Maritime Raiding, Irregular Warfare, and the Early American Navy

Benjamin Armstrong

University of Oklahoma Press | Norman

Publication of this book is made possible through the generosity of Edith Kinney Gaylord.

Library of Congress Cataloging-in-Publication Data

Names: Armstrong, Benjamin, author.
Title: Small boats and daring men : maritime raiding, irregular warfare, and the early American
　　Navy / Benjamin F. Armstrong.
Description: Norman : University of Oklahoma Press, [2019] | Series: Campaigns and
　　commanders ; volume 66 | Includes bibliographical references and index.
Identifiers: LCCN 2018039814 | ISBN 978-0-8061-6282-9 (hardcover : alk. paper)
Subjects: LCSH: United States. Navy—History—18th century. | United States. Navy—
　　History—19th century. | United States. Marine Corps—History—19th century. | United
　　States—History, Naval—To 1900. | Naval tactics. | Irregular warfare—United States. | Raids
　　(Military science)
Classification: LCC E182 .A76 2019 | DDC 359.00973—dc23
LC record available at https://lccn.loc.gov/2018039814

*Small Boats and Daring Men: Maritime Raiding, Irregular Warfare, and the Early American
Navy* is Volume 66 in the Campaigns and Commanders series.

The paper in this book meets the guidelines for permanence and durability of the Committee on
Production Guidelines for Book Longevity of the Council on Library Resources, Inc. ∞

Contents

Illustrations

Figures

Map

Acknowledgments

No voyage of this distance or operation of this scope can be, or should be, charted alone. There are many who helped make this book possible. In some ways, the story begins and ends in Annapolis, Maryland. It was inside the high granite walls of the U.S. Naval Academy that I first experienced the academic study of history. My parents instilled in me a love of books and of water, if not necessarily the sea, and my high school history teacher, Joe Thomas, prepared me well (insisting in our European History class that I read an obscure book about European navies by some old dead guy named Mahan). But it was in the classrooms in Sampson Hall that my love of history really took root. In particular, I owe much to Professor Craig Symonds, who grabbed my attention as a young midshipman, continued to inspire me as I read his books while deployed far from home, and became an academic mentor when I returned to Annapolis.

The faculty of the Norwich University military history graduate program introduced me to doing my own academically viable scholarship and understanding the discipline of history. The project that eventually expanded into this book began at King's College London, where I read for my PhD with Professor Andrew Lambert, Laughton Chair in Naval History. Andrew offered himself as a sharp intellectual challenger and a supportive and flexible supervisor throughout my studies. He has my eternal gratitude. My secondary supervisor John Bew, the faculty involved in the Laughton Naval History Unit at King's, Marcus Faulkner, Alan James, and Alessio Patalano, and my examiners John Hattendorf and John Beeler have been obliging and encouraging, and they continue to be my guides as I navigate occasional shoals. King's War Studies Department also offered me a group of fellow voyagers who have made an enormous difference, including Nick Prime, Alan Anderson, Andrew Breer, and many others who are helping to develop the future of naval history.

Outside of faculties and the ivory tower, this project and my efforts to grow as an officer and historian have had many supporters. Dr. Jerry Hendrix, who reached out to me when he was a commander and I was a lieutenant, has provided constant mentorship and advice on navigating the stormy seas of scholarly pursuit while serving as an operational naval officer. Some of my superiors in the Navy have been very encouraging of my work. In particular I owe thanks to Shawn Bailey, then a commander, who sent me off to induction in London from our helicopter hangar in Norfolk, and to Capt. Pete Brennan and the men and women in the Office of the Secretary of the Navy, who helped while I served on staff duty. Thanks also belong to the pilots and sailors of the "Kraken," who tolerated my ramblings at the wardroom table while at sea and kept me alive in the air. Some of my superiors and peers have not been encouraging; in fact, quite the opposite. They may still believe, as Capt. Frank Ramsey once wrote on Alfred Thayer Mahan's fitness report, that "it is not the business of naval officers to write books." But I won't list them. There are many names missing from this brief list of friends and supporters. The Society for the Repression of Ignorant Assumption knows that they have my thanks and they know who they are.

Adam Kane has not only been a magnificent editor for me, but a great sounding board and a good friend. His team at the University of Oklahoma Press has been efficient and helpful in guiding this book through the publishing process. Series editor Greg Urwin and the peer reviewers made this book far stronger than when it started. I have the staff of many institutions, like the University of Oklahoma Press, to thank for helping me develop my work, including many supportive journal editors and my friends at the U.S. Naval Institute, and the Society of the Cincinnati, which helped with a research fellowship.

And now it has come full circle. Many of the professors with whom I studied as a midshipman are now my colleagues as I learn the ropes in the History Department at the U.S. Naval Academy. They have been an infinite source of encouragement and advice. My fellow military professors, both former and current, are magnificent examples and are always available to share their wisdom. I admit that I am often in need of it. I am also thankful for the department's junior scholars, who have accepted an inexperienced academic, but an overly experienced officer, into the fold, and helped to develop a sense of community that creates great historians and teachers.

Finally, the person who has done the most to support me, deserves the greatest appreciation, and has my undying love: my wife, Charity. Her sacrifices would require a book-length explanation of their own. From the moment that I brought up the idea of striking out on my own path to study history, instead of getting the MBA or computer science degree the Navy wanted me to have, she has been there for me. Deployments and the challenges of life married to a naval aviator aside, she has somehow dealt with my absentmindedness, piles of books and unsorted papers, and constant research and study trips with a grace that I do not deserve. There is so much more to say, but she is also my best editor and would admonish me that I have made my point and I need to get on with it. This book is dedicated to her, and all the sacrifices she has made for me.

SMALL BOATS AND DARING MEN

INTRODUCTION

On a clear and cool April morning in 1906, a crisp wind blew down the Severn River as the Naval Academy in Annapolis came to life for a day of pomp and circumstance. The midshipmen donned their dress uniform, and Lt. Cdr. John A. Hoogewerff took charge of their formation while the band and marine battalion finished preparations for the ceremonies. After lunch, with everything arranged for the arrival of the official party, Academy superintendent Rear Adm. James Sands led a convoy of new automobiles to the Annapolis train station where they met President Theodore Roosevelt and his delegation from Washington, D.C. The day marked the 126th anniversary of the Revolutionary War battle between HMS *Drake* and the American sloop of war *Ranger*, captained by John Paul Jones. For the United States Navy, it was a perfect day to honor the memory and legacy of one of its great captains. For Paul Jones, whose remains had been discovered in an unmarked Parisian grave only two years before, it marked not only a return home but also a notable step toward his eventual enshrinement as the father of the U.S. Navy.[1]

The ceremony honoring John Paul Jones was a fitting backdrop for President Roosevelt, who was advocating for the construction of a fleet commensurate with the United States' newfound responsibilities and opportunities in the wake of its victory over Spain in 1898. Paul Jones became one of the symbols Roosevelt used to illustrate the potential of American naval greatness, in his politically charged effort to increase the size and power of America's sea services. Later that day in the Academy's armory, the president delivered a speech to the invited dignitaries and the gathered midshipmen. He challenged opposition leaders in Washington, D.C., to try to block his effort to build a bigger navy when he said, "Those of you in public life have a moral right to be here ... only if you are prepared to do your part in building up the Navy."[2]

The proof of America's need for a large blue-water force, Roosevelt declared, began with John Paul Jones. The president, an accomplished naval historian in his own right, noted Paul Jones's victory over *Drake* as well as HMS *Serapis*, open-water battles fought by warships with their great guns. He then connected those victories to the battles of the War of 1812 in which American ships in the blue water, or squadrons on the Great Lakes, defeated the Royal Navy. Those successes had not been enough, however, according to Roosevelt. They were insufficient because there were too few of them, and there were too few because there were not enough large warships to fight the big battles necessary for overall success. Roosevelt argued that great officers victorious in small battles were important, but they had to be a part of a large and powerful navy and marine corps that could win a war. In doing so, he transformed John Paul Jones's re-interment at Annapolis into a defining symbol of his own agenda: the United States needed a powerful blue-water battle fleet as it entered the twentieth century.[3]

Roosevelt's appropriation of the memory of John Paul Jones distorted the actual history of the Revolutionary War hero. As a result, the president abandoned a balanced view of Jones's legacy. While the victories over *Drake* and *Serapis* certainly had great value in the Scots-American captain's own memory and in his own explanations of his operational achievements, they were by no means the pinnacle and exclusive accomplishments of his career. Instead, John Paul Jones's vision for the Continental Navy, as well as the operations he conducted, included a great deal of maritime raiding and what today would be called naval irregular warfare. Small-boat raids, cutting-out expeditions, and small-unit amphibious landings played a central part in both how John Paul Jones conceptualized American naval warfare and in how he executed his own operations. From raiding the Solway Coast and cutting out enemy ships in the bays of Nova Scotia, to a planned amphibious prison break from a Canadian island and small-unit raiding operations on the Gulf Coast, Paul Jones was about much more than blue-water ship duels. Looking across American naval history during the Age of Sail, it becomes clear that the conduct of these kinds of operations was a near-constant endeavor. While Roosevelt gave the U.S. Navy a father with eyes fixed on the blue-water battle, the historical record shows that naval raiding and irregular warfare represent an essential, if unstudied, theme in the history of American sea power.

The prevailing historical narrative of the early United States Navy traces the growth of the blue-water force that would become the large and powerful transoceanic fleet of the second half of the twentieth century. From Paul Jones against *Serapis* and the frigate duels of the War of 1812, through Manila Bay to Midway, the ship-on-ship or squadron-on-squadron engagements dominate the approach of most naval history. Although in the early American experience these tended to be part of commerce raiding, which later grew and developed into sea control, historians and officers both privilege them as the form of naval activity most appropriate for study. Regardless of whether pursuing what strategists have called *guerre de course*, or attacks on enemy commerce and trade, or executing *guerre d'escadre*, or naval war via fleet and warship battles, these conventional operations dominate the history books. This is particularly true of the history of combat operations, but it is also true of how the administration, training, and construction of naval forces are studied.[4] Alfred Thayer Mahan wrote that the best use of a navy is to find and defeat an opponent's fleet, and much of the history that Americans write about their navy is refracted through a Mahanian prism.[5] That narrative covers an important aspect of the U.S. Navy's past, yet it gives an incomplete view. From the earliest days of the Republic, the U.S. Navy was involved in operations other than fleet, squadron, or ship-on-ship engagements and it conducted these other missions on a global scale regardless of the contemporary size or shape of the U.S. fleet. This history forms an important, but almost entirely overlooked, element of the American naval tradition.

This book examines early American naval history through the lens of *guerre de razzia*, or war by raiding, and the irregular operations required to conduct this type of naval warfare. Traditional maritime thinkers have divided naval strategy into the bifurcated structure focused on guerre d'escadre, and guerre de course.[6] But, this classical structure overlooks the multitude of naval operations that do not fit into the dogma of oceanic battles between massed fleets and attacks on an enemy's trade and supply lines. The history shows that the early American navy and marine corps's conduct of naval raiding and irregular operations was a fundamental part of how American commanders executed their orders in both war and peace. The events in the following chapters indicate that there were, in fact, common operational and tactical patterns of naval irregular warfare that demonstrate the evolution of guerre de razzia as an operational and strategic concept in parallel to the development of American fleet operations and commerce raiding.

The eight episodes in the following chapters highlight themes that identify naval irregular warfare as a coherent subfield of operational naval history. This adds important balance to—and expansion of—existing scholarship on the individual events but also advances a historically based inquiry into irregular operations. The themes that emerge involve the people who led and participated in the events, similarities in the vessels and weapons used, and the contributions of partnerships or relationships with organizations and individuals outside American naval forces. In these three areas of analysis, demonstrable patterns emerge that contribute to an understanding of how naval raiding and irregular warfare operations developed through the period and help explain their success or failure. In addition to illustrating the central place of irregular operations in American naval history during the Age of Sail, they offer examples that buttress the identification of guerre de razzia as an element of naval warfare.

———————————

As Sir James Cable pointed out in his works on gunboat diplomacy, once scholars move outside the bounds of conventional naval warfare based on fleet and warship engagements, definitions become the subject of debate and sometimes even distraction.[7] Since Cable embarked on his study in the 1970s, an even greater variety of definitions related to military concepts and terms—such as asymmetric warfare, military operations other than war, hybrid war, and irregular warfare— have muddied the waters as scholars and practitioners adopt multiple definitions.[8] Some scholars have labelled the kinds of maritime operations detailed in the following chapters as "nontraditional."[9] However, that label is inappropriate since many navies have a long history and tradition of these types of operations, even if they have not been widely studied. Likewise, other scholars may attempt to define these as "littoral operations" because they occur in the shallows and inshore waters of the world. While it is a geographically accurate label, Nelson's victory at the Battle of the Nile and the fight between the *Monitor* and the *Merrimac* at the Battle of Hampton Roads also occurred in the littoral region. Those engagements, and others, were most certainly in keeping with the traditions of fleet battle and ship-against-ship combat. Therefore, labelling the operations studied in the following chapters as littoral, while accurate, is insufficient.

Just in these two examples, the difficulty in labels and categorization becomes clear. This book casts the net widely, using naval irregular warfare as a label describing the types of missions and tactics that are often employed in maritime raiding. Naval irregular warfare includes two groups of combat operations: peacetime maritime security operations and wartime raiding operations. The focus here is specifically on small-craft and small-unit missions and the tactics employed in the execution of guerre de razzia. During wartime, naval irregular warfare includes maritime raiding operations conducted outside of conventional and historiographically dominant ship-on-ship, squadron-on-squadron, or fleet-on-fleet engagements. This includes small-unit amphibious raids, cutting-out expeditions, small-boat attacks, and other similar tactical missions. These often take the form of what Vice Adm. Philip Colomb described as "harrying" or "cross-ravaging" operations.[10] The targets may be ashore, or equally likely afloat. It does not include the more traditional naval operations of privateers, since they tended to be open-water operations and ship-versus-ship engagements and were a part of classically defined guerre de course. There are examples of missions conducted as part of coastal defense in the following chapters, but the work does not examine the gunboat navy of the Jeffersonian era. Other historians have examined the short-lived gunboat navy extensively. Their studies indicate that while the swarming small boats envisioned by the gunboat proponents might be described as asymmetrical, they were not irregular. At their most basic, the gunboat operating concepts and tactics still envisioned vessels cannonading each other until victory in a conventional surface engagement, even if one side was made up of small combatants while the other was large fleet warships.[11] Likewise, the episodes in this book do not examine large-scale amphibious assaults to invade another nation, or power projection in the form of bombardment, because they remain a part of conventional naval strategy.[12]

During peacetime, navies conduct irregular operations as the combat portion of maritime security operations. This includes missions conducted to protect resources and sovereignty; to counter maritime criminal activity including smuggling, piracy, and terrorism; and to ensure open and free use of the sea.[13] Such constabulary duties have remained relatively constant for naval forces, and within maritime security operations the episodes in the following chapters focus on the use of combat, or the threat of combat, to achieve those ends. This is consistent with the use of the word "warfare" in the label naval irregular warfare. There is not an examination of humanitarian assistance missions or pure naval diplomacy, despite the fact that some recent naval writing has labelled them irregular warfare. While modern theories and doctrines inform this definition of maritime security operations, it also aligns with early American concepts of the navy's role as a protector of commerce and defender of global interests.[14] It suggests that guerre de razzia and irregular operations extend beyond strictly wartime naval strategy and have a place in wider maritime strategies pursued during relative peace.

Besides debates over categorization, there is admittedly a second complication with using naval irregular warfare as a doctrinal description for the missions and tactics involved in maritime raiding: American naval officers, sailors, and marines in the Age of Sail did not use the phrase. The frequency of these missions, combined with the understanding that they required different skills, training, and experience beyond dueling broadsides, suggests that naval men of the era would likely have understood the term once described, but it is still the result of modern classification. These kinds of missions were seen as a central part of naval warfare in the nineteenth century, as demonstrated by the U.S. Naval Academy's naval history examination in the 1881–82 academic year. Of twelve possible essay questions, which covered the first century of the U.S. Navy and Marine Corps and included the battles of the Civil War and frigate duels of 1812, two questions covered episodes specifically examined in this book, the 1820s counterpiracy campaign in the Caribbean and the frigate *Potomac*'s mission to

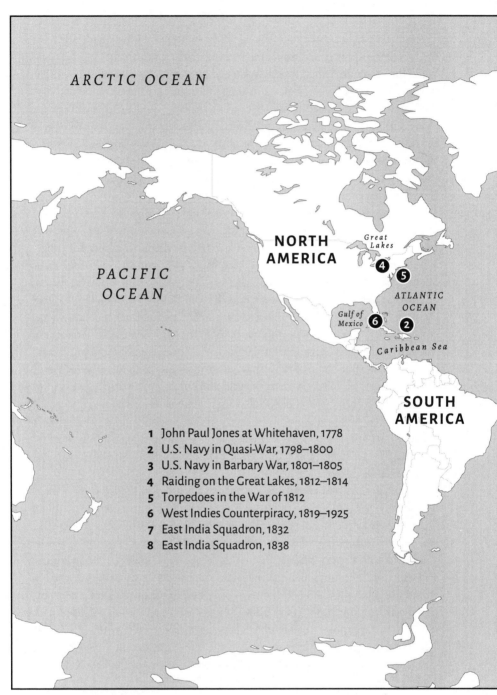

1 John Paul Jones at Whitehaven, 1778
2 U.S. Navy in Quasi-War, 1798–1800
3 U.S. Navy in Barbary War, 1801–1805
4 Raiding on the Great Lakes, 1812–1814
5 Torpedoes in the War of 1812
6 West Indies Counterpiracy, 1819–1925
7 East India Squadron, 1832
8 East India Squadron, 1838

Global Operations of the Early U.S. Navy.
Map by Erin Greb Cartography.

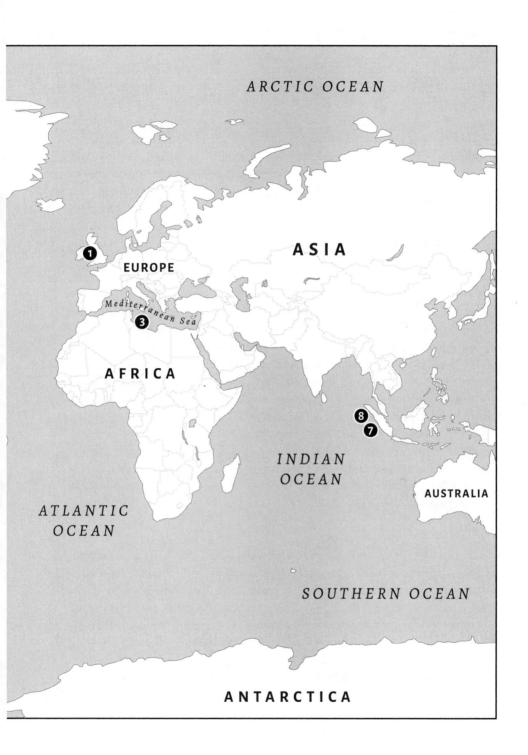

ARCTIC OCEAN

ASIA

EUROPE

Mediterranean Sea

AFRICA

*INDIAN
OCEAN*

AUSTRALIA

*ATLANTIC
OCEAN*

SOUTHERN OCEAN

ANTARCTICA

9

Sumatra.[15] This suggests that the professors and officers at Annapolis considered the understanding of these missions as essential professional knowledge for an aspiring officer candidate.

Officers in the early American naval service also did not use or learn doctrinal concepts in a way familiar to later sailors or marines. Professional learning was far more ad hoc and was based on personal mentorship and individual teaching. Early American officers would not have used the phrases guerre de course or guerre d'escadre either, despite the fact that they have become de rigueur for naval historians when describing the era.[16] While many modern historians may lament the use of a phrase like irregular warfare, ignoring how modern military thinkers and practitioners describe conflict risks allowing the study of history to be equated with simple antiquarianism.[17] It is the first step down a path that dismisses history as a less valuable contribution to military professionalism and knowledge. Even if officers of the era did not label the episodes in the following chapters as irregular warfare, they recognized that the missions studied here were something different from the blue-water squadron and warship actions, and still an important part of naval affairs, the same way they should be recognized today.

In an effort to examine the subject of raiding and naval irregular warfare with width and depth, and within the wider context of the events, this book does not examine the subject in an exhaustive way. Documenting every example of maritime raiding and naval irregular warfare in the Age of Sail would necessitate an encyclopedic book, and would give up the ability to examine the subject in detail. Instead, this book takes an episodic approach and examines eight examples of irregular missions or irregular campaigns across the Age of Sail period. This provides a deeper examination of the archival record, and the opportunity to determine the similarities and the differences between the examples studied.

The first episode this book examines is the cruise of the Continental Navy's sloop of war *Ranger* under the command of John Paul Jones, including the ship's expeditionary raiding operations on the Scottish coast. James C. Bradford began his initial examination of guerre de razzia in Paul Jones's correspondence, but this chapter examines the operational and tactical history to determine how exactly the early American captain executed his raiding plans. Paul Jones conducted a cruise on board *Ranger*, which included both the conventional naval missions of commerce raiding and engaging enemy warships, and also irregular raiding missions. The raids on Whitehaven and the Earl of Selkirk's estate by sailors and marines marked the first examples of naval irregular warfare far from American shores, providing a baseline from which to begin a detailed examination of further examples.

The second chapter explores the mixing of wartime maritime raiding and peacetime maritime security operations in the United States's first "undeclared" war: the Quasi-War with France, 1798–1800. The history of the Quasi-War introduces how maritime security operations like counterpiracy mixed with wartime missions like small amphibious raids in a hybrid conflict. American naval forces only engaged the French navy in the Caribbean. But they also had to protect American merchant shipping from privateers and from pirate attacks, as well as understand the differences between the two. Simultaneously, they faced the diplomatic challenges of operating in a small area with many national allegiances.

The third chapter follows a similar examination, again of an "undeclared" war, which followed almost immediately on the heels of the Quasi-War: the First Barbary War, 1801–5. The Barbary War offers a similar operational dynamic, but one with a national force of corsairs rather than pure pirates as popular historians assert, and a coastal geography and territory rather than archipelagic operations. The history from the Barbary conflict focuses on the small ketch *Intrepid*, from its initial capture as the merchant vessel *Mastico* through its irregular operations, which offer a demonstration of how small combatants play important roles in naval raiding.

Two chapters center on the War of 1812. The first covers raiding and cutting-out expeditions on Lake Ontario and the Great Lakes region. These operations targeted both British warships and logistics in the campaigns from 1812 through 1814. The frontier naval arms race, which developed on the Great Lakes, was dependent on the flow of materiel from the eastern seaboard in the case of the U.S. Navy, and from Britain and along the St. Lawrence River in the case of the Royal Navy. In the early years of the war some of the local watermen on Lake Ontario volunteered for naval service and received warrants as sailing masters. They joined with naval officers and sailors to begin small-boat raids and gunboat operations on the lakes. The squadron engagements and shipbuilding race have overshadowed these operations conducted by small boats and masked their role in the conflict on the lakes.

The second War of 1812 chapter focuses on the Atlantic seaboard and surveys the nexus of the U.S. Navy and civilian partisans in torpedo operations against the Royal Navy's blockading force. Despite the dominance of the blue-water ship duels and the Great Lakes squadron engagements in the War of 1812 scholarship, after the British blockade descended on the American coast a less well-known campaign began. It was a campaign of naval irregular warfare and partisan operations against the blockading fleet, a campaign by a weaker power against the global naval hegemon over access to the defender's coast, and a campaign with particular relevance to challenges faced by the twenty-first century U.S. Navy. In

March of 1813, the U.S. Congress passed the Act to Encourage the Destruction of the Armed Vessels of War of the Enemy (known as the torpedo act, authorizing prize money for anyone, civilian partisans or military service members, who sank a British warship in American waters. Inspired by the torpedo designs of inventor Robert Fulton, American watermen in Long Island Sound and Chesapeake Bay developed innovative tactics and technologies to strike at the British blockade. The techniques of the torpedo boat attacks and mining operations link these new methods of naval irregular warfare with previous and future examples.

The fifth chapter examines the counterpiracy and maritime security operations in the Caribbean Sea and the Gulf of Mexico during the 1820s. Piracy was a growth industry in the Caribbean during the first decades of the nineteenth century. In 1821, attacks on American merchant shipping had risen to the point that Congress called for action, and President James Monroe formed the West Indies Squadron to protect American shipping. Between the initial deployment, in 1821, and 1825, when attacks had declined to almost zero, three different commodores commanded the squadron in its maritime security operations. Coordination with the Royal Navy, and mixed results in attempts to cooperate with Spanish authorities in Cuba and Puerto Rico, offer examples of the interaction between naval irregular warfare and naval diplomacy and partnership. These peacetime campaigns provided numerous specific events to study in a lengthy period, including changes in leadership, strategy, and operational design that illustrate the character of naval irregular warfare.

The sixth and seventh chapters investigate U.S. naval operations on the coast of Sumatra in the 1830s. Two separate piracy incidents in the same area on the northwest coast drew a reaction from American warships and their sailors and marines. In 1831, Commo. John Downes and the frigate *Potomac* responded to an attack on the New England spice trader *Friendship*, which resulted in a landing operation and battle at the village of Kuala Batu. In 1838, another attack on a spice trader drew a response from Commo. George Read and his ships, the frigate *Columbia* and corvette *John Adams*. The attacks, which instigated an American naval response, were very similar. However, how the individual commanders carried out their responses differed significantly, and this offers an opportunity to compare the connection between diplomacy and combat, as well as leadership decisions, in naval irregular warfare.

A more complete and rigorous view of the U.S. Navy and Marine Corps's past must include study of the maritime raiding and naval irregular missions that make up a large proportion of American naval operations on a day-to-day, year-to-year basis. Ships on the blue water of the world's oceans, and their development as a conventional fighting force capable of victory in ship-, squadron-, and fleet-level engagements, are certainly a vital part of the operational history of the U.S. Navy. Guerre d'escadre and guerre de course are central elements, and may even be the dominant elements, of the naval past. They are not, however, the only elements. Maritime raiding and naval irregular warfare, far from being a small number of isolated or insignificant occurrences during the Age of Sail, represent an important thread of American naval history and therefore American sea power. This commonality demonstrates the existence of guerre de razzia as a third element of sea power, and a concept deserving greater study. Even as the field of naval history has expanded alongside the "new military history," incorporating social, economic, and political history among other areas, the focus has tended to remain on the development of forces that would participate in the decisive sea battle. Historical engagement with missions like naval raiding operations or counterpiracy patrols has lacked comparison or connection to larger maritime or naval themes. Guerre de course and guerre d'escadre continue to dominate the way historians and strategists see naval affairs, despite the fact that navies have not used these doctrines operationally since 1945.

Cooperative operations to maintain maritime security, the use of small-unit raiding and targeted irregular operations to achieve military objectives in war, the execution of irregular operations against nonstate adversaries, and other elements that play central roles in the history examined, are also important parts of naval operations in the modern day.[18] From the raids in the Quasi-War and the First Barbary War and the counterpiracy operations at Sumatra, to the special-operations raids launched from ships against twenty-first-century terror cells and counterpiracy off Somalia, naval irregular warfare has been a fundamental part of American naval affairs in the past and the present, and will be in the future. Naval irregular warfare is not irregular at all; it has always been a critical aspect of American sea power.

CHAPTER 1

John Paul Jones and the Birth of American Naval Irregular Warfare

His Majesty's schooner *Gaspee* jerked to a stop in what Lieutenant William Dudingston thought was clear water. The schooner, on revenue and security patrols in the Narragansett Bay of Rhode Island, was following a suspicious packet sloop named *Hanah* toward Providence. But *Gaspee* had not been up the Providence River before, and her sailors and commander did not know the hazards of the sandbar off Namquid Point, a sandbar that *Hanah*'s skipper Thomas Lindsey had purposely sailed close to in the hope that the falling tide might snare the British warship. It was June of 1772, and the relationship between the watermen of Rhode Island and the Royal Navy was at a low point after months of seizures, inspections, and tax collection.[1]

Dudingston and his men tried to break *Gaspee* free from the sand, but to no avail. With the tide continuing to fall, and the sun setting, it would be hours until the water rose enough to free the schooner. The British sailors estimated that around 0200, in the early morning hours, the ship would lift off the bar. Dudingston ordered his watch to set a lookout and sent the rest of his crew below to get some sleep. He retired to his cabin and changed into his bedshirt to catch some needed rest himself.

Just after midnight, as the crew slept, Seaman Bartholomew Cheever spotted something in the darkness. He first thought he had simply seen the moonlight play off nearby rocks. However, a few moments later he looked again and the rocks appeared to be moving. He shouted a warning but received no response across the water. He hailed a second time, again with no answer. The lone watchstander headed aft and ducked into his skipper's cabin, waking Dudingston and William Dickinson, the midshipman who was *Gaspee*'s second in command. As the seaman went back on deck, the lieutenant grabbed his officer's sword and followed him out into the night in his bedclothes.[2]

Back on deck, the British made out a pair of boats approaching *Gaspee* from the north, the lead group of what would be a whole swarm. Dudingston shouted toward them, demanding that they identify themselves. A voice rose out of the night from Abraham Whipple, identifying himself as the sheriff of Kent County, Rhode Island, and announced that the boats were approaching to serve a warrant and arrest Dudingston. He was charged with overstepping his authority in pursuit of smugglers along Rhode Island's shores. Dudingston ordered them to depart and to return to discuss their issues at a more appropriate hour. He sent Dickinson to unlock the small-arms locker as more boats appeared and continued to approach *Gaspee*.

Whipple again shouted his intention to arrest the lieutenant, encouraging the men who rowed the boats to pull on toward the warship. Dudingston ordered Cheever to fire a warning shot with his musket. But the seaman's flint had become wet and the lock snapped without firing. As the boats rushed the final yards toward the sides of the ship, the lieutenant shouted down the hatch to awaken his crew and order them on deck. He rushed for the starboard bow, where the first boat approached, and swung his sword toward the ghostly men who were attempting to board his ship out of the night.[3]

A musket shot rang out from the boats. Dudingston was hit in the left arm, the ball continued and ripped through his groin, and he fell back onto the deck in a pool of his own blood. The boarders swarmed over the schooner's low bulwarks, armed with axe handles and wooden staves, and beat the British crew back below decks, guarding the hatch to ensure they could not come out. The raiders thronged around Dudingston and Dickinson, menacing in the moonlight and threatening to beat them both further unless they begged for their lives. But Abraham Whipple and John Brown stopped them, the ship secured for the moment. Dudingston struck a deal with the colonials, he would order his men to surrender if they promised nobody else would be hurt. The men from Providence agreed.[4]

Eight boats had gathered at Fenner's Wharf in Providence when *Hanah* arrived, telling the tale of *Gaspee*'s grounding. Whipple, a former privateer skipper in the French and Indian Wars, and Brown had gathered and armed crews with muskets, knives, and clubs. *Gaspee* had been preying on Rhode Island merchantmen, stopping and delaying their voyages, seizing their cargo with what they considered questionable cause. They had had enough, and the local merchants had sworn out a complaint against Dudingston and a judge issued an arrest warrant. The eight boats had rowed the five miles south to *Gaspee* to launch their maritime raid.[5]

The British sailors surrendered, under the lieutenant's orders, and the colonials began binding their hands and placing them into the boats. The Rhode Islanders treated Dudingston's wounds and carried him off the ship as well. With the schooner empty, they set combustibles and lit them before scrambling for the boats left alongside. Pulling hard for the shore, the raiders and their captives watched as *Gaspee* burst into flames, the fire climbing into the rigging and light-ing up the night sky. They abandoned the sailors on the shore near Pawtucket, and the boats rowed back north toward home as the British warship burned through the night. Nothing but the hulk below the waterline remained.[6]

American naval and maritime history has included irregular operations from the earliest days of the American Revolution. In the years before the exchange of gunfire at Lexington and Concord, traditionally seen as the start of the armed struggle, the colonists and British authorities had already experienced conflict and violence at sea. With the burning of the revenue privateer *Liberty* in Newport Harbor in 1769, and then the violent destruction of His Majesty's schooner *Gaspee* in 1772, American revolutionaries adopted maritime irregular warfare long before the conflict spread ashore.[7]

Following the outbreak of the war on land, irregular warfare continued to play a part in American maritime operations. Most histories of the Continental Navy, however, have tended to focus on conventional blue-water operations like warship duels and attacks on enemy shipping, whether by Continental Navy ships or privateers. Some historians have examined the building and administration of the conventional naval force.[8] Capt. John Paul Jones, held up by Roosevelt as the "father" of the American Navy, offers an example that demonstrates the understudied place of naval irregular warfare in the operational history of American navies. The biographies and histories of Paul Jones focus on the climactic blue-water battle off Flamborough Head between his ship, the East-Indiaman-turned-Continental-frigate *Bonhomme Richard*, and HMS *Serapis*. A conventional frigate duel, the violence and glory of that event provide the ultimate example of Paul Jones's naval skill and accomplishment.

That was what Roosevelt targeted when he appropriated Paul Jones's legacy. However, the captain himself appeared to have looked at things differently. In his memoir to King Louis XVI, he spent as much time discussing, and appeared to have taken as much pride in, the maritime raiding operations he conducted as in the blue-water victories. The wider use of raiding and naval irregular warfare in the American Revolution, from the multiple raids on the Bahamas and the whaleboat men of American coastal waters, to the Mississippi River raiding of the Willing Expedition, requires further study. But, because of Roosevelt's focus on Paul Jones, and the general acceptance of his characterization of the Scots-American captain, this chapter will focus explicitly on the raiding and irregular operations conducted by John Paul Jones.[9]

Paul Jones considered irregular operations to be as important a part of naval warfare as conventional sea battles and saw it as a central element of American sea power. He certainly played an influential role in establishing the foundations of early American naval affairs regardless of debate over the "father" label.[10] But, as recounted in the introduction, Paul Jones received an important place in the blue-water culture of the twentieth-century U.S. Navy when President Roosevelt invoked his memory. Because of his importance both in his own era and in later naval affairs, taking a closer look at the operational history of Paul Jones offers a straight-forward, and appropriate, starting point for the examination of the American history of naval irregular warfare in the Age of Sail.[11]

Born John Paul, the fifth child of a gardener from the Scottish Lowlands, as a young man the future American captain apprenticed to a shipowner in White-haven on the Solway Coast. Putting to sea as at a young age, he had risen through the ranks of the merchant trade rather quickly, taking command of his first ship by the age of twenty-one. Trouble with a murder charge in the West Indies contributed to his decision to immigrate to America and join his brother, a tailor in Fredericksburg, Virginia. After settling in as an American, he assumed the new surname "Jones."[12]

When the American Revolution broke out Paul Jones volunteered for the Continental Navy. Surprisingly he turned down command of the 12-gun sloop *Providence* in order to take a lieutenant's commission on board *Alfred*. He believed command in the merchant trade, even successful command and skill as a mariner, did not automatically prepare an officer for combat duty in a navy.[13] Sailing in March of 1776 with the squadron under Commo. Esek Hopkins, *Alfred* participated in the American navy's first offensive action of the war, a raid on New Providence in the Bahamas.

Ignoring his orders to cruise the Chesapeake and the southern coast to defend the rebellious colonies from the Royal Navy, Hopkins instead decided to sail farther south in search of desperately needed military stores. In a stroke of luck, the squadron sailed into the Bahamas after British commanders had removed the company of regulars from the British Army's 14th Regiment and the Royal Navy's patrol ship *Savage* from Nassau for more pressing orders. Paul Jones, having sailed the Bahamas on a merchantman, helped serve as a pilot for the expedition, and the first American raid on foreign shores landed with almost no resistance. The landing force of sailors and Continental Marines captured both Forts Nassau and Montague without a fight, and they took custody of several dozen cannon and military supplies. But the raid was only a partial success, as local British residents helped an enterprising captain named William Chambers drop his cargo of lumber overboard and carry 162 large barrels of desperately needed gunpowder away in the night before the Americans could capture it.[14]

After the raid on New Providence, the squadron returned north to Rhode Island to deliver the captured ordnance. As they approached Long Island, the ships ran in with the frigate HMS *Glasgow*, but much to Paul Jones's embarrassment the entire squadron failed to defeat the lone British ship. Paul Jones learned a great deal from this first cruise. He seemed to have a better sense of the military necessities and tactical understanding required for successful raiding, and after seeing how the captains had handled their ships against *Glasgow*, he gained even more confidence in his own abilities, and doubts about the other officers in the nascent navy. He gladly accepted command when Hopkins offered it following the Nassau raid.[15]

His first ship was the 70-foot sloop of war *Providence*, in which he sailed on 21 August 1776 for a successful commerce raiding cruise to Canadian waters. He and his crew captured sixteen vessels, destroying half and bringing the others in as prizes. Paul Jones and his men, however, used more than conventional open-water pursuit and blue-water battle to achieve their ends. Even though the ship was small, he sailed with a detachment of thirteen Continental Marines under the command of their captain, Matthew Parke, as well as seventeen soldiers from the Rhode Island Brigade. These thirty extra men offered Paul Jones expanded opportunities to attack British interests. Stopping in Canso Bay, Nova Scotia, to collect water and firewood, they discovered a number of British ships, which they attacked and burned, and a small fast-sailing schooner named *Defiance*, which they captured and manned as an auxiliary to work in concert with *Providence*. A few local fishermen came on board and volunteered to join the crew, bringing with them intelligence on British merchant ships and fishermen anchored farther along the coast around Île Madame.[16]

Paul Jones sailed *Providence* off the harbors of Arichat and Petit de Grat and loaded the new auxiliary and one of his boats with sailors and marines. He sent one raiding party into each harbor on cutting-out expeditions. They caught their targets completely by surprise and in both locations, the British fishermen and merchant sailors surrendered en masse as the Americans swarmed over their sides from the small boats. The operation was such a success that Paul Jones did not have enough men to man the prizes and guard the prisoners at the same time. As a compromise solution, he turned two of the captured vessels into cartels to carry prisoners, all Jersey fishermen, home as parolees. Then he refitted and manned the other four ships as prizes and returned to Narragansett Bay.[17]

After Paul Jones was lauded in the newspapers for his success on board *Providence*, Commo. Esek Hopkins re-assigned him as the captain of *Alfred*, the ship in which he served as first lieutenant on his first deployment. He had the brig *Hampden* and *Providence* placed under his command and orders for another cruise to Canadian waters. This mission, however, had a new element for Paul Jones. Instead of just attacking the Newfoundland fishery and the ships, which carried coal from Cape Breton Island to the mainland, he was ordered to land on Cape Breton and raid the mines to free American prisoners used as laborers.[18] Commodore Hopkins had intelligence that there were twenty ships in the harbor, part of the fleet that supplied coal to the British forces in New York, as well as one hundred American prisoners in the mines. Protecting these British assets were the Royal Navy sloop *Savage* and the brig *Dawsons*, but Hopkins's intelligence indicated only one of them was present at any given time.[19] The interesting and unique orders excited Paul Jones. He wrote to his friend Robert Morris, who was a member of the Marine Committee of the Continental Congress as well as his regular correspondent: "[A]ll my humanity was awakened and called up to action by this laudable proposal."[20] The orders, along with his experience in the assault at New Providence and the cutting-out at Ile Madame, began his early thinking on the strategic and operational aspects of maritime raiding. This included not only a desire to strike at British interests beyond American shores but operations to free British-held prisoners.

The expedition, however, never reached Cape Breton. Paul Jones was unable to man all three ships fully, which led him to delay his departure and forced him to consolidate his manpower. He sailed with only two ships: *Alfred* and *Providence* finally departed on 1 November.[21] Because they had not sailed earlier, the Atlantic winter set in and Paul Jones and his ships encountered "contrary winds and stormy weather."[22] Despite this, they still had a productive cruise. After taking several prizes and weathering a gale, which separated *Alfred* and *Providence*, Paul Jones found himself again off Canso Bay. He manned his boats and sent a raiding party ashore, which burned a transport full of provisions and torched a warehouse containing the region's whale oil and the fishing equipment in the town. They also captured another small, fast-sailing schooner to use for future raids. Over the next four days, *Alfred* captured three transports, a merchant ship, and 140 prisoners. The Americans learned from their prisoners that British warships had sailed for the area in search of them. They also learned that the 44-gun British ship *Roebuck* was anchored in Spanish River on Cape Breton, to defend the harbor, which made any attempt to free the American prisoners almost impossible. Overwhelmed by the need to man his prizes Paul Jones elected to take his small fleet of captured vessels and run for port, reaching Boston on 15 January.[23]

The late departure and winter weather had kept Paul Jones from launching his operations to free the prisoners at Cape Breton, but he had still taken the opportunity to raid ashore. Prior to departing on the cruise, he was already considering the role of far seas raiding for the American cause. In another letter to Robert Morris, he suggested a winter cruise to the coast of Africa with *Alfred* and two of the smaller Continental ships. He even considered the possibility of attacking and capturing the island of St. Helena to serve as a rendezvous and base of operations from which he would prey on the slavers of the British Africa trade. He pointed out to Morris that the small ships he was requesting did not do well on the American coast in the winter, and African waters would provide ample hunting ground where they would succeed in "not leaving them [the British] a mast standing on that coast."[24]

However, after the return from his early winter cruise, Paul Jones and Morris changed their plan to something much more daring. While Africa could provide a lucrative set of targets, an expedition to British West Florida offered even greater opportunity. Morris outlined how a few small sloops, responsible for patrolling the four hundred miles of shoreline from the Mississippi to Apalachicola Rivers, only lightly defended the colony on the Gulf Coast. There were also reports the garrison at Pensacola had a significant cache of artillery pieces, which would make an enormous contribution to the American cause, but few troops to protect them. An attack at Pensacola, commerce raiding off the mouth of the Mississippi River, and a possible maritime raid ashore at St. Augustine on the return voyage looked like a dramatic but achievable plan.[25]

Morris suggested raids ashore on British colonies would have a more substantial impact than mere commerce raiding: "[Commerce raiding] alone is an object worthy of your attention, but as I have said before, disturbing their settlements and spreading alarms, showing and keeping up a spirit of enterprize, that will oblige them to defend their extensive possessions at all points is of infinitely more consequence to the United States than all the plunder that can be taken. If they divide their Force we shall have elbow room."[26] The two men believed that, beyond the tactical implications of maritime raiding, strategically it would force the British to redistribute their force for the protection of the empire, which would relieve some of the pressure on American shores. With Paul Jones's agreement to the plan for Florida, Morris sent a letter from the Marine Committee to Hopkins instructing him to detach *Alfred*, *Columbus*, the brig *Cabot*, *Hampden*, and *Providence* under Paul Jones's command for what the latter described as an "enterprize of the greatest importance to America."[27] After the plan to raid the mines at Cape Breton fell through, Morris continued to influence Paul Jones on the importance of irregular operations for American naval affairs.

But Commodore Hopkins did not like the idea of detaching so many of his ships for his subordinate to command on a secret expedition. He had other plans for the vessels and severe manpower problems to confront. Also, the relationship between Paul Jones and his commodore had begun to unravel. At Hopkins's behest, the Marine Committee of the Continental Congress ordered Paul Jones to turn over *Alfred* to another captain and to return to *Providence*, which he saw as a demotion, and it led to a growing feud. Paul Jones's and Morris's scheme died from both lack of resources and lack of support caused by his rising conflict with Hopkins.[28] Instead, the Marine Committee changed Paul Jones's orders again, this time sending him to take command of a frigate, which the Americans were obtaining in Europe. The Committee attempted to arrange for him to sail on board the French ship *Amphitheatre* that had arrived with crucial military stores for Gen. George Washington's army. Paul Jones was in Boston, awaiting the final details of his voyage to France, when the plan fell through.

In November 1776, the Continental Congress in Philadelphia had authorized the construction of a new group of ships to reinforce the nascent Continental Navy. The law authorized three 74-gun ships of the line, five 36-gun frigates, and an 18-gun brig as the newest warships of the Continental Navy. John Langdon, the continental agent at Portsmouth, New Hampshire, who ran a shipyard at Badger's Island on the Piscataqua River, received the contract for the brig.[29] Upon receiving the order, Langdon wrote to Alexander Hamilton at Philadelphia and told him he "immediately ordered one of my Master builders out of the Yard, with a Gang of Hands into the woods to Cut and procure Timber for the Vessell, which I am order'd to build [*sic*]." Langdon planned to build the continental ship along the same lines as a privateer he had constructed, but he suggested to the committee that he should rig the ship as a sloop of war, which he believed would be easier to handle in open ocean sailing and heavy winds.[30] Using the rig out of the captured British merchantman *Royal Exchange*, which had a similar size and type of rigging, would make things easier for the shipbuilder. It would be cheaper as well, since as the naval agent he already had custody of the prize.[31] When the plan for Paul Jones to sail to France on board *Amphitheatre* collapsed, the Marine Committee ordered him to take command of the sloop in Portsmouth, now named *Ranger*, and to have the ship "equipped, officered and manned as well and as soon as possible."[32]

Paul Jones arrived in Portsmouth and set about fitting out the ship for sea. His dealings with Langdon while preparing the ship continually frustrated him, despite what he considered clear orders from the Marine Committee to expedite his departure. From the availability of sailcloth to a less than adequate supply of rum, the two men struggled over the fitting out of the vessel and its costs.[33] The Marine Committee gave Langdon and local committee member William Whipple responsibility for helping Jones appoint the officers for the ship. The result was a wardroom composed of the favorite sons of Portsmouth with Thomas Simpson, Langdon's brother-in-law, appointed first lieutenant. Second lieutenant Elijah Hall, lieutenant of marines Samuel Wallingford, and the surgeon Ezra Green were all local acquaintances of either Langdon or Whipple.[34]

In the fall of 1777, *Ranger* finally filled her compliment, mostly from Portsmouth locals, and Paul Jones was ready to sail for France. He completed his wardroom by appointing Matthew Parke, his marine commander from *Providence*, as captain of marines. *Ranger* waited out a series of gales off the New Hampshire coast and then set sail for France on 1 November 1777. When she sailed, the 116-foot sloop of war carried a crew of 140 and only eighteen 6-pound guns, instead of the twenty-four 6-pounders originally procured, because Jones feared the ship was top-heavy.

The early weeks of the Atlantic crossing were uneventful for *Ranger* and the crew. Lookouts spotted few sails, and Paul Jones used the time to instruct and drill his crew. He concluded they "have not formerly been conversant in the management of ships of war" but saw the potential in such a crew because, "they pay more attention to instruction than conceited fellows who think themselves two [*sic*] knowing to be taught." His initial impression of the crew was positive.[35] The first excitement for the crew came with dawn on the morning of 19 November. Lookouts spotted a sail under the lee quarter. *Ranger* bore off the wind to allow the ship to pass and discovered a large ship, which showed no intention of slowing or heaving to for an inspection. Paul Jones ordered the ship to quarters and the men prepared for action as they gave chase, but lost their quarry in the night.[36] Approaching the Azores, *Ranger* entered the shipping lanes frequented by British vessels. Day and night, as lookouts spotted ships on the horizon, Paul Jones beat his men to quarters and *Ranger* began to develop the habits of a warship. On the crossing, they boarded several ships and took two prizes.[37] On 2 December, after weathering a storm in the previous two days, the lookouts spotted the French coast. At two in the afternoon, a French pilot came on board and *Ranger* entered the river Loire. By four she was anchored, and Paul Jones took the cutter into Nantes to begin making arrangements for the ship.[38]

Paul Jones was well satisfied with the way his crew had performed during the crossing. He wrote to John Wendell that the voyage had given him "the most agreeable proofs of the active spirit of my officers and men."[39] The crew proved themselves capable mariners, making it through the gale of 28–30 November without much struggle, but he also was beginning to have hope for their preparation for naval operations.[40] The drilling of the crew had gone well in the early part of the voyage, and once they passed the Azores the men had responded well to the drums calling them to quarters. In a letter to William Whipple, Paul Jones pointed out that he had been able to exercise the crew at night, the most difficult time for combat operations, and "their behavior was entirely to my satisfaction."[41] *Ranger*, however, was another matter. The Atlantic crossing and the storm revealed a number of issues. Paul Jones's initial suspicion, that by borrowing the rig from *Royal Exchange* the builders had over-sparred the sloop, turned out to be accurate. He also felt *Ranger* was still improperly balanced, even having left some of his cannon behind, and needed more ballast. He ordered iron ballast to be laid under the stores in the ship's hold and had the crew begin shortening the masts and re-rigging the ship.[42] The captain spent his time in France between the ship, where he oversaw her re-sparring, ballasting, and the sewing of new sails, and Paris, where he lobbied for a larger and more powerful ship.

By 16 January 1778, the American commissioners in Paris had given up their attempts to obtain a frigate for John Paul Jones. They issued him new orders: "[A]fter equipping the *Ranger* in the best manner for the cruise you propose, that you proceed with her in the manner you shall judge best for distressing the enemies of the United States, by sea or otherwise, consistent with the laws of war, and the terms of your commission. . . . We rely on your ability, as well as your zeal to serve the United States, and therefore do not give you particular instructions as to your operations."[43] His hopes of coming to France to take command of a frigate and to become commodore of a squadron in European waters dashed, Paul Jones intended to take full advantage of the open-ended nature of his orders.

The correspondence does not spell out explicitly the intention to conduct a cruise that included raiding ashore, yet it was clear as early as January 1778 that he was considering it. He appears to have discussed it with some of the Continental commissioners while in Paris, because they mention it twice in correspondence with him. On 15 January 1778, the Marine Committee pledged they would help the crew of *Ranger* obtain rewards for "special" missions where they are unable to carry away prizes, which would include raiding ashore.[44] In the orders they issued to Paul Jones on 16 January they explicitly instruct, if "you make an attempt on the coast of Great Britain, we advise you not to return immediately into the ports of France, unless forced by stress of weather."[45] It is apparent he planned a cruise that included the traditional naval missions of commerce raiding and engaging the enemy's warships, but also irregular missions like amphibious raiding ashore. His experiences at New Providence and in commanding raids from *Providence* and *Alfred*, combined with the planning for the African and West Florida expeditions, led him in one clear direction: the shores and harbors of the British Isles.

On 13 February 1778, *Ranger* sailed from the Loire and headed northwest along the coast of Brittany to Quiberon Bay. Arriving that evening as the wind began to pick up, the Americans anchored short of a French squadron, which occupied a part of the bay. In Quiberon Bay, Paul Jones had the first significant indication that there was some strife among his junior officers. The men from Portsmouth, First Lieutenant Simpson, Second Lieutenant Hall, and Sailing Master David Cullam, signed a letter expressing their concern over Captain Parke's presence on board *Ranger*. They pointed out that a ship of less than twenty guns did not rate a captain of marines. They already had a lieutenant of marines in their friend from Portsmouth, Samuel Wallingford. The reason for their objection, however, was not propriety or naval tradition but instead the fact that Captain Parke was diluting their share of the prize money, both from the two ships already captured but also from future prizes. They asked Paul Jones to consider sending Parke ashore or to the frigate *Deane*, which was in France and would soon sail for America and which rated an officer of his seniority. The marine officer, and trusted subordinate of Paul Jones, left *Ranger* of his own choice to join the crew of *Deane*.[46]

Several weeks later, on 10 April 1778, Paul Jones and his crew of New Englanders departed Quiberon Bay. The French frigate *Fortuna* escorted *Ranger* to sea, since the British were watching the port. On the morning of 13 April, *Fortuna* parted company with *Ranger*, "sooner than Capt. Jones expected."[47] The next day, after some early morning fog, the Americans spotted and chased their first sail. Around eight in the morning, they pulled abeam the brig and spoke with the captain. Paul Jones determined the ship was sailing from Ostend on the Flemish coast to Galloway in Ireland, making it a legitimate prize. He ordered the crew taken on board *Ranger*, and since the vessel had a cargo of flaxseed and appeared to have little value he ordered the prize sunk.[48] The captain did not want to reduce his crew so early in the cruise by manning a prize of such low value, though sinking their very first prize had an effect on the crew's morale and expectation of prize money. On the morning of 17 April, as the sun rose with cloudy skies, they again spotted a number of sail and gave chase. By eight in the morning, they pulled alongside *Lord Chatham* out of London, carrying a mixed cargo and a significant amount of English porter. *Lord Chatham* was a larger ship than the brig they had sunk, with a more valuable cargo. In sight of the British coast, Paul Jones ordered her master, William Starhorn, and crew taken prisoner. He appointed Master's Mate John Seward as prize master and sent him with a crew of eight men to take the ship and the prisoners to the Admiralty Court at Brest.[49]

Sailing the western coast of Scotland, Paul Jones raised the idea of a raid ashore for the first time. With the familiar waters of Whitehaven nearby, he saw an opportunity to make the attack on British soil he had been considering. The response of the crew was less enthusiastic, and Surgeon Green was purported to have pointed out: "[N]othing could be got by burning poor people's property."[50] The disagreement between the New Englanders, who, it had become apparent, thought of themselves as commerce-raiders, and their captain, who sought glory and a militarily significant cruise, came to a head for the first time. The chief motivation for most of *Ranger*'s sailors, some of whom the naval officers had recruited away from privateering billets, was prize money. A mutinous plot developed, which Paul Jones discovered and foiled by surprising one of the leaders with a pistol to the forehead. Between the discontent on board the ship and the winds and seas that made a small-boat operation unfavorable, Paul Jones abandoned his plan. For the moment.[51]

The next evening, east of the Isle of Man near Ramsey Bay, *Ranger*'s lookouts spotted a schooner and they hoisted the top gallant yard and gave chase. It was not until five the next morning, with a fresh wind under a cloudy sky, that they approached what they identified as a "King's cutter." After a week, at sea *Ranger* had found its first British naval vessel and the crew beat to quarters. The British revenue wherry *Hussar*, out of Whitehaven and commanded by Captain James Gurley, turned toward *Ranger* when the British skipper decided the American ship looked suspicious. With his ship disguised as a merchantman, Paul Jones intended to allow the cutter to come alongside for a customs inspection before hauling up the American colors and capturing the crew as they attempted to board.[52]

Something gave away the Americans, however, and the British ship turned to run before coming alongside. Paul Jones ordered his men to open fire. *Ranger*'s marines fired their small arms at the vessel as she outmaneuvered the larger American ship. Paul Jones was able to get his broadside into position, holing the *Hussar*'s mainsail twice and scoring a hit on her stern. Unfortunately, as Paul Jones later reported, "this vessel out-sailed the *Ranger*, and got clear, inspite [*sic*] of a severe cannonade." Green later blamed the escape on Paul Jones, because "had the Captain have [*sic*] permitted the Marines to fire on them when they first came under our lee quarter, might have taken her with great ease."[53]

The rest of 19 and 20 May, *Ranger* tacked between the west coast of Scotland and the east coast of Ireland. She ran in with a number of coasters and, after taking off the crews, she scuttled a sloop out of Dublin and a schooner carrying a cargo of oats and barley. The schooner revealed there were ten or twelve ships at anchor in Loughryan on the Scottish coast. Paul Jones considered a raid on the harbor, like his attacks at Canso Bay, but the winds were against him and he elected to "abandon my project."[54] The evening of 20 May, a fishing boat with a crew of six from Belfast came alongside and *Ranger* detained them and took their craft in tow. The Irish fishermen offered intelligence that the British 20-gun sloop HMS *Drake* lay just north of their position at Carrickfergus outside Belfast.[55]

Paul Jones ordered the ship to quarters and told his crew he intended to sail into Belfast Loch and cut out the enemy warship. Green recorded serious discord among the crew in his diary, writing, "the wind blowing fresh and the people unwilling [to] undertake it we stood off and on till midnight."[56] Livid at another near mutiny, Jones adjusted his plan and worked out a scheme to surprise the British and take the enemy ship intact. At the darkest part of the night, and after the winds subsided so the crew agreed, *Ranger* sailed into the harbor and prepared to anchor in the vicinity of the British ship. The captain's plan was to get close enough to foul *Drake*'s anchor with his own. This would cause *Ranger* to drift against *Drake*'s side and just forward, but would look to the world like an accident or poor seamanship. In that position, musket fire from the American marines could rake the deck. Grappling hooks would go over the side and hold the two ships together so his crew could board and take the enemy before the British knew what happened.[57]

After midnight *Ranger* headed up the channel, disguised again as a merchantman attempting to escape the heavy seas and winds. Paul Jones brought his ship alongside *Drake*, but the crew did not let go the anchor when commanded, whether due to the difficult weather or, as Paul Jones later insisted, the drunkenness of the mate. The anchoring maneuver did not work and left *Ranger* a half cable's length from *Drake*. Out of position to board and unprepared to attempt a cannonade, but still undetected, Paul Jones cut his cable, while trying to make it look like an accident in the rising seas, and slipped back out of the harbor.[58]

Overnight *Ranger* weathered a heavy gale and spent the next two days running under close-reefed sails in high seas and winds. The Americans skirted around the Isle of Man, attempting to keep clear of the Irish Sea's hazards, as the storm swept them from Ireland to Scotland. After spending all night at quarters in the attempt on *Drake*, and then sailing through the gale, Green reported, "Our people were very much fatigued."[59] The weather began to let up on the afternoon of 22 April, and the Americans found themselves off the harbor of Whitehaven, Scotland.[60] American ships had cruised the British home waters in search of prizes and easy merchant prey before. In 1777, *Lexington* and *Reprisal* made similar cruises.[61] In the first weeks of her cruise, *Ranger* had done the same: carrying out conventional naval missions by attacking enemy merchantmen and engaging enemy warships. However, on the morning of 23 April 1778 the crew of *Ranger* launched something different, something more in keeping with Paul Jones's operational and strategic views for the Continental Navy: a pre-dawn small-boat raid on the harbor at Whitehaven.

Paul Jones was raised on the Solway Coast of Scotland, and Whitehaven was the home-port for the first ship he served on board as an apprentice. He knew the waters and the harbor well and could serve as his own pilot.[62] As *Ranger* approached the coast on 22 April, a fresh blanket of snow lay ashore as far as the eye could see. The ship sailed in close to the town, but as the evening approached the winds died and the ship's progress slowed. Farther out from the harbor than he had hoped, Paul Jones assembled his raiding party and embarked them in the boats. He divided thirty-one sailors and marines between the two boats, with Paul Jones himself in command of the first and Lt. Samuel Wallingford and Midn. Benjamin Hill in the second. Around midnight they began to row for shore.[63] The crews pulled at their oars for more than three hours before they reached the outer pier. Despite the danger of carrying out their attack with the approach of sunrise, Paul Jones ordered Wallingford and his party into the northern branch of the harbor where between seventy and one hundred small, shallow draft ships and boats lay "on the hard," aground in the tidal area. The town had no sense of their danger and had no guards posted on the wharfs or the pier. The vessels on the hard were empty as well, except for a few sleeping boys, with the tide slowly beginning to rise.[64] Paul Jones took his own boat to the south side and scaled the wall into the main fort, which guarded the town from a seaborne attack.

Once inside, the raiders discovered the fort's handful of watchmen huddled in their guard house, taking refuge from the cold, and blockaded them inside without firing a shot. While the main group from the boat spiked the guns, Paul Jones and a sailor named Joseph Green proceeded to the southern battery and spiked those cannon as well. They disabled all of the town's defensive armament, more than thirty 32- and 42-pound guns, and then returned to their boat and crossed back to the northern side of the harbor.[65] Paul Jones expected to find ships ablaze and the raiders under Wallingford hard at work spreading the inferno. The men had armed themselves with cutlasses and pistols, and they were equipped with a number of different "combustibles," to fire the grounded fleet.[66] However, instead of finding the harbor descending into a conflagration, Paul Jones found the second boat and its men "in some confusion."[67]

The candles the men carried in lanterns from the ship had gone out, and they needed to rekindle the flames to start the ships afire. While the party carried flint and steel, they struggled to relight the flame needed to renew their attack. Paul Jones posted pickets and sent his men into a watch house at the end of the pier, separated from the rest of the town. A small fire burned in the hearth of the house: the watchman's family huddled in bed, terrified by the armed men who burst into their home. The Americans relit their lanterns and rushed back into the night. In the confusion over relighting the candles, however, they missed the fact that one of their own had slipped away. David Freeman, also known to Ranger's crew as David Smith, who later claimed to be a loyal Irishman who had shipped on board Ranger to find a way home, broke into the nearest public house and began knocking on doors and raising the alarm.[68]

The sun continued to rise, and as the Americans returned to their task in the harbor the alarm spread through the town. Paul Jones selected a large ship near the center of the anchored and grounded fleet and set torches and combustibles ablaze in her hold. He selected well: the ship Thomson was one of the newer and more valuable vessels in the Whitehaven fleet and considered "one of the finest ever built here."[69] The Americans found a barrel of tar and rolled it into the flames, hoping it would set fire to other ships since time was running out. Approaching an hour after dawn, with the sun fully above the horizon, Paul Jones reported, "the inhabitants were appearing by the thousands and individuals ran hastily towards us. I stood between them and the ship on fire with my pistol in my hand and ordered them to retire which they did with precipitation."[70]

The morning daylight revealed how small the party of American raiders was. Paul Jones ordered his men back in the boats. They released most of the hostages they had taken, with only room for three in the boats. The captain took a moment to enjoy the look of amazement on the faces of the town's leading "Eminences" before shoving off from the pier and rowing hard for the safety of *Ranger*. The townspeople rushed for the fort but discovered the raiders had spiked all their cannon. They managed to fire a single shot, perhaps from a gun overlooked by the Americans, but the ball fell well short of the escaping boats. The raiders fired their pistols in a return salute. *Ranger*, with a freshening wind, sailed in close to recover the boats and then set back out into the Irish Sea.[71]

The raid came off without a single casualty. David Freeman had deserted, but the Americans had taken three prisoners in return. However, Paul Jones was disappointed in the results. The men watched as smoke rose from the town, but it was not enough to indicate the fire was spreading. The people of Whitehaven turned to the firefighting effort with expected zeal and, while *Thomson* was a total loss, they managed to put out the fire before it spread and engulfed the harbor.[72] Paul Jones lamented to the American commissioners in his report that he had not started the operation earlier: "[H]ad it been possible to have landed a few hours sooner my success would have been complete: not a single ship out of more than two hundred could possibly have escaped; and all the world would not have been able to save the town."[73] The leading men of Whitehaven agreed, saying in their report to London: "[H]ad two or three ships been set on fire; in that case all the ships in the harbor must have been consumed, with some parts of the town."[74]

Ranger sailed north across Solway Firth. The raiders were exhausted. They had stood by for the attempted cutting-out of *Drake*, weathered the gale the evening of 21 and 22 April, and then assaulted Whitehaven in the night. Paul Jones, however, was a flurry of activity. As the ship approached Kirkcudbright Bay, rather than tack away from shore, *Ranger* came to and Paul Jones ordered the cutter into the water. The American ship lay off St. Mary's Isle and the estate of the Earl of Selkirk. Lord Dunbar Douglas was a minor member of the British aristocracy and was not an active or particularly political member of the nobility.[75] However, he was an important figure in the mind of Paul Jones, who knew him from having grown up in the region. American prisoners held by the British remained in Paul Jones's thinking, likely along with the memory of the aborted rescue attempt at Cape Breton Island. He reasoned that if he could capture Lord Selkirk, and hold him hostage, he and the commissioners in Paris could negotiate a large prisoner exchange, which could free Americans held on both sides of the Atlantic.[76]

As in the attack on Whitehaven, Paul Jones's personal knowledge of the Scottish waters paid dividends. With Sailing Master Cullam, Lieutenant Wallingford, and nine fresh members of the crew, the captain took personal command of the cutter and piloted it up the narrow and winding channel that led to the estate. The landing party put ashore and began moving toward the estate. They came across a group of locals, who did not recognize them or the purpose of their visit, and who told Paul Jones that the earl was in London and only his wife and her friends were at home. His plan to capture the earl foiled, the captain resolved to head back for the cutter but his men disagreed. The marginal success in Whitehaven, and the growing discontent among the crew about the number of prizes Paul Jones had sunk rather than taken in for prize money, led the rest of the landing party to insist they receive something for the effort of landing at St. Mary's. Paul Jones agreed to allow Cullam and Wallingford to take the men up to the estate to demand a ransom of the household silver. He ordered the sailors not to enter the house or commit any kind of attack or pillage. Unwilling to participate in such a low endeavor himself, he then returned to the cutter to await their return.[77]

Cullam and Wallingford obeyed the captain's orders to the letter. When they reached the front door of the estate, they ordered the men to take up a watch around the house and only the officers entered. Met by Lady Selkirk, they demanded the household silver. An account from the scene, later reported in the London newspapers, expressed surprise at both the "mixture of rudeness and civility" in the treatment by the Americans. Also unexpected was the fact they demanded nothing besides the silver, not even the jewelry Lady Selkirk or her guests were wearing, "which is odd."[78] In an almost anticlimax to the raid, once they had the household plate in hand, the landing party walked back down the hill to the cutter where they met their captain and rowed back out to *Ranger*.[79]

In light winds and under hazy skies, *Ranger* crossed back to the western shore of the Irish Sea during the night and the crew received some much-needed rest. The news of the attack on Whitehaven, and the intelligence of the "American Privateer," spread to the Royal Navy's ships in the region. Captain John Gell of the frigate HMS *Thetis* put to sea from his anchorage in the River Clyde. He knew the frigate *Boston* was also in the vicinity and wrote to the Admiralty that if he came across the other ship he would give Captain William Dudingston orders to assist in the search. He set aside his other missions for the moment.[80] The Admiralty also ordered the 20-gun ship *Heart of Oak* to join the pursuit.[81] By six o'clock on the morning of 24 April, the Americans were off Belfast, once again within sight of *Drake*. Paul Jones beat his men to quarters with the intention of sailing into the harbor to take on the British warship. However, his officers and men again disagreed with him. Fighting the ship in the anchorage at Carrickfergus could easily mean there would be no prize, even if *Ranger* was victorious, since it would be difficult to get a damaged vessel out of the anchorage. Instead, Paul Jones had them clear the decks and make the ship look like a merchant sailor again. Commander George Burdon, captain of *Drake*, had received word of the attack on Whitehaven. At nine o'clock, the lookout on board *Ranger* reported *Drake* was getting under way from the anchorage.[82]

The wind blew from the east, making it a long transit of tacking upwind for *Drake* to exit the harbor and reach the open water where *Ranger* waited for battle. In the five hours *Ranger* waited, Burdon sent a boat ahead with a midshipman and seven sailors to check on the ship, which appeared to be a merchantman, and to impress extra sailors for *Drake*'s crew. As the midshipman surveyed *Ranger* through his glass, Paul Jones kept his stern to the approaching boat, and his men out of sight, which convinced the approaching party that this was a merchant vessel. The British boat came alongside and the midshipman climbed on board to begin the impressment.[83] With the British officer standing before him, Paul Jones revealed himself and his crew and took the eight men captive. The capture, according to the captain, had "an exhilarating effect on my crew," and they went back to quarters and prepared for the coming battle.[84]

The civilians ashore realized *Ranger* was the American raider, instead of just a merchantman trying to stay out of the way, and they lit alarm fires. Five small craft, which had put out with *Drake* full of civilian onlookers, began to think better of their choice and drifted back toward the beach. After a long sail against the wind and the tides, around six o'clock *Drake* cleared the shoals and came on toward *Ranger*. Paul Jones led the ship away from the Irish coast toward deeper water then waited for the British.[85] As evening approached and *Drake* closed on *Ranger*, the British hoisted their colors. According to Paul Jones's report to the commissioners in Paris, "at the same instant the American stars were displayed on board the *Ranger*." He expected the formalities were over and the battle was ready to commence. However, *Drake* continued toward the Americans and hailed them, asking what ship they were. Paul Jones, surprised the British seemed to be unprepared or confused by the situation, had Mr. Cullam reply "the American Continental Ship *Ranger*—that we waited for them and desired that they would come on."[86]

The British did not respond but continued closing from *Ranger*'s stern. Paul Jones waited until they were within pistol shot and ordered the helm up to wear the ship and bring his guns to bear. He did not want to wait any longer, with sunset only an hour away, and ordered his men to open fire. The first broadside tore across *Drake*'s decks and into her rigging, and the battle commenced. Paul Jones related, "the action was warm, close, and obstinate."[87] In the light winds, the two ships engaged in the first battle between an American and British warship in British waters. An hour into the engagement one of *Ranger*'s marines fired into the mass of men on *Drake*'s quarterdeck. The musket ball struck Commander Burdon in the head, dropping him to the deck. His lieutenant, a volunteer named William Dobbs, was struck down moments later. The Americans shot away *Drake*'s ensign twice and there was no flag to strike. In the carnage on board the British ship, the frantic waving of a hat on the quarterdeck accompanied a desperate shout for quarter, and the British surrendered.[88]

An American party boarded *Drake* and began rounding up prisoners and assessing the damage. Burdon's wounds were fatal; there was nothing Surgeon Green could do for him. He died within hours of the end of the battle. Likewise, Dobbs passed within days of the fight. The crew buried both men at sea with the honors required for naval officers.[89] The Royal Navy lost forty-two men, killed and wounded in the battle. Lieutenant Wallingford and Seaman John Dougal died in the fighting on *Ranger*, and six men were wounded. The rigging and sails on *Drake* were "entirely cut to pieces."[90] Nearly all her masts and yards were damaged and her hull as well. *Ranger* weathered the battle much better, with most of the damage to her low and running rigging, which proved easy to repair. All day on Saturday, 25 April, the Americans worked to repair both ships as they drifted south along the coast of Ireland.[91]

The crew completed the heavy repairs on 26 April, and lookouts spotted a brig approaching. The ship was unaware a battle had taken place and sailed close aboard to inquire about the two ships lashed together and drifting in the open water. Paul Jones reported he "was obliged to bring her too [sic]." *Patience* was out of Whitehaven and, in repayment for the master's concern about the welfare of fellow mariners, or more likely out of fear that he would report their location, Paul Jones took the ship as a prize and put a crew on board. With the three ships ready to get under sail, the captain decided to head north. The shortest route back to the safety of French waters lay to the south, but the winds were contrary and sailing north and circumnavigating Ireland would take the Americans farther away from the Royal Navy's bases.

After drifting southward, *Ranger* and her prizes again passed the scene of their battle off Belfast Loch as they headed north. Paul Jones decided to put ashore the Irish fishermen whom the Americans had captured the week before. The men had provided valuable intelligence for the first attempted attack on *Drake* and the raid at Whitehaven, and seemed to have no qualms about helping the Americans. Paul Jones referred to them as "the honest Fishermen," and because *Ranger* had lost their boat in the stormy seas, he gave them a boat from one of the captured ships and paid them for the loss of their equipment. The Irishmen agreed to carry two grievously wounded British sailors ashore, along with one of

Drake's sails, to prove what would be a surprising story to the authorities when they arrived with the injured men. As they sailed off for Belfast, the fishermen appeared grateful and acknowledged the Americans with three cheers.[92]

Sailing north toward the open ocean, the British revenue wherry *Cumbras* spotted and shadowed *Ranger* and *Drake*. Captain James Crawford, commanding *Cumbras*, gave up the chase to report their position to HMS *Thetis*, and the Americans' apparent decision to head north. However, "thick and rainy" weather settled over the region, and Captain John Gell of *Thetis* asked Crawford to ensure he reported *Drake's* loss and the Americans' possible course northward to the Admiralty. He feared the difficulty he would have finding them in the weather and knew most of the Royal Navy's ships were probably searching to the south in the Irish Sea.[93]

As Paul Jones's little squadron headed into the Atlantic the Royal Navy lost them. After an uneventful trip, on 7 May the ships arrived at Brest after a cruise of twenty-eight days.[94] The crew had few prizes to show for it, but they did have more than two hundred prisoners that Paul Jones intended to ransom to the British, in an effort to free American privateers held in the infamous Mill prison and elsewhere. The American commissioners in Paris responded to Paul Jones's report of his actions, congratulating him on "the honour you have acquired by your Conduct and Bravery in taking one of the King's Ships."[95]

The attacks on British soil caused an outcry in the press and among the population, both along the coasts and in London. Following the attack on Whitehaven, the town's leaders dispatched express riders along the coast and to the major cities. The local magistrate ordered all strangers and out-of-town guests rounded up for questioning. The leading men of the town recommended to their peers on the coast that they do the same.[96] False reports began to spread. On 9 May, the *London Advertiser* reported the 20-gun *Heart of Oak* had been attacked and taken by the Americans, a false story based in the mania developing on the British coast. Newspapers reported other erroneous reports of American "privateers," and the inhabitants of the coastal seaports were in "daily expectation of being plundered."[97] Insurance rates for the trade across the Irish Sea reportedly quadrupled in the week following *Ranger's* attacks.[98]

Local authorities called out the militia, and the authorities in London ordered regiments of dragoons to encamp on the coast to be ready to respond to further attacks. Because of poor manning of the Royal Navy's ships in home waters the Admiralty increased impressments by a thousand men to fill out their complements.[99] Local reports stated that "there are the greatest preparations making, every one fitting up and repairing their old rusty guns and swords, making of balls, etc."[100] Officials set about trying to rectify the generally defenseless state of the British coast, including the purchase and arming of several floating batteries in harbors around Great Britain. False alarms continued to sound for weeks, each time resulting in the calling out of the militia and the setting of guards.[101]

A month after *Ranger* had returned to France, the coast was still awash with worry over impending American attacks. In the *St. James Chronicle*, an anonymous writer published a poem entitled "The Naval Slumber" which mocked the Crown and the Royal Navy for their inability to protect the British people.[102] The "Pirate Paul Jones" became a fearsome threat in the minds of many British subjects but also resulted in a real political problem for the government. The *London Evening Post* published a sarcastic letter purported to be from Paul Jones to Lord Sandwich, First Lord of the Admiralty, to thank him for "permitting me, for so long a time, to seize plunder, and carry off the vessels of the merchants, in the British and Irish Seas."[103] Newspapers opposed to the government wrote opinion columns pointing out that British behavior in America, burning towns and destroying the homes of innocents, were just as responsible for American attacks on the British Isles as any sort of ill character displayed by the "pirates," driving discussion of the righteousness of the British side of the war.[104]

The raids by Paul Jones on the coast of Scotland had an impact on the British psyche. Ten years later, in his travelogue of parts of Great Britain, the writer William Gilpin could not pass the southwest coast of Scotland without mentioning and relating the story of Paul Jones's "depredation," casting him as "daring, and enterprizing," but all the more dangerous because he was "low, illiberal, and unprincipled." Even a hundred years after the landing at Whitehaven, naval historian Sir John Knox Laughton wrote that the "Pirate" Paul Jones was still a threat in "northern nurseries" and had taken on a mythic place alongside the Jabberwock and the Bandersnatch. The relatively defenseless state of much of the English and Scottish coast, even at the turn of the next century, continued to concern Laughton.[105] Amphibious invasion and occupation is a great threat in maritime warfare, but irregular operations and raiding, particularly as part of a larger and balanced strategy, can have a psychological, political, and sometimes economic effect (though not economic in this case because of failure to burn the ships) on littoral populations that far outweighs the resources required for such missions.

While the cruise of *Ranger* was only a month long, it demonstrates a number of principal elements that play a role in naval irregular warfare. As with all military operations, the quality of the leadership displayed by commanders and officers is important. But in irregular warfare, that leadership requires a different perspective. The cruise of *Ranger* illustrates the relationship between irregular operations and conventional naval missions, including conventional blue-water combat. The episode also demonstrates the interaction between naval forces and maritime populations. This results in the need for a unique kind of intelligence and knowledge, different from blue-water operations. Gathering this kind of information commonly points to the importance of partnership with local allies.

Naval irregular warfare requires a particular quality of leadership: senior officers who understand the diplomatic and strategic implications of their actions and empowered and aggressive junior officers. Preparation for conventional naval warfare tends to diminish the need for either of these, despite their importance in peacetime or irregular situations. Instead, senior officers with an extreme focus on fleet maneuvering, and seamanship-based operational experience, are required for those who aspire to command in large conventional ship battles. A larger strategic or diplomatic understanding of the ramifications of their decisions, beyond the tactical, does not tend to be required of individual captains. In the Age of Sail, this meant senior officers who aspired to command squadrons or fleets focused on mastery of command and control through signals, understanding and maximizing the capabilities of their particular vessel, and a view toward wartime service on the blue water. In the case of junior officers, exacting execution of technical skill and procedural compliance were required for conventional naval warfare expertise. For juniors this meant the ability to manage a division on the gun decks, ensuring rapid and exact compliance with orders and procedure.

On board *Ranger*, the irregular mission benefited from a senior officer who clearly met both the need for tactical and conventional ability but also a greater strategic and diplomatic understanding. Paul Jones's pursuit of naval irregular warfare throughout the American Revolution, through both executed and planned raiding missions, demonstrated his ability to rise to a higher level of understanding. Even after his disappointment in the tactical execution in the harbor at Whitehaven, he still knew his larger goal had been accomplished. He wrote to the American commissioners: "What was done however is sufficient to shew [sic], that not all their boasted Navy can protect their own Coasts—and that Scenes [sic] of distress which they have occasioned in America may soon be brought home to their own doors."[106]

However, *Ranger* lacked the aggressive junior officers necessary for complete success in their irregular missions. Neither first lieutenant Thomas Simpson nor second lieutenant Elijah Hall was even willing to volunteer for the expedition into Whitehaven. They remained focused on the desire for prizes and prize money, and had little apparent interest in how to advance the American cause. Samuel Wallingford, the lieutenant of marines, showed his own lack of leadership skill, which compounded his lack of experience. Accusations later arose that, rather than trying to burn ships in the harbor, he instead led his men into a pub where they proceeded to get drunk. There is no indication of this in the contemporary reports, however, and it is likely based on dubious claims in the British press or the fact that the first place the deserter went was a pub. Even discounting the story of drunkenness as an example of rationalizing the incomplete result, Wallingford still showed little initiative and struggled to accomplish his task and lead his men. In retrospect, Paul Jones's decision to send Captain Parke home prior to sailing on the cruise was a mistake. He later wrote to Congress, "[H]ad [my former officers on board *Providence* and *Alfred*] been with me in *Ranger*, 250 or 300 sail . . . would have been laid to ashes."[107]

The importance of using irregular warfare as part of a broader concept of naval operations was evident in the open-water engagements *Ranger* experienced. She took part in numerous conventional blue-water operations; including commerce raiding, which resulted in the prizes the crew craved. She also engaged the enemy's warships, including the unsuccessful engagement with *Hussar*. Without the ability to conduct conventional naval warfare, and succeed in the ultimate conventional battle against *Drake*, *Ranger*'s cruise might have been considered a disappointment and disaster. Two decades prior to the cruise of *Ranger*, the Royal Navy experienced a similar challenge when the French privateer François Thurot raided the north coast of Ireland with a small squadron of ships. But history remembers that episode as an example of the Royal Navy's successful defense of the British Isles, because after his raid Thurot's ships were defeated and captured, and Thurot himself killed, in a battle with the frigates *Aeolus*, *Pallas*, and *Brilliant*.[108] Understanding the balance between irregular and conventional naval warfare, and the need to prepare and train for both, was key to the success of Paul Jones's cruise in British waters.

Finally, local knowledge and proper intelligence was also central to the success of *Ranger*'s operations. Paul Jones selected Whitehaven for his attack for a simple reason: it was his hometown. He had grown up sailing and fishing the waters as a boy. He could lead his boats into the harbor on a dark night knowing exactly where the piers were located and the arrangement of the batteries. His local knowledge was vital at St. Mary's Isle as well, since the approaching channel was long, narrow, and unmarked in the late eighteenth century. But Paul Jones also gathered knowledge from other sources. Use of local guides and allies can be an effective method to gain the intelligence needed for irregular missions. It was not only the captain's personal knowledge that was important, but also the Irish fishermen taken on board *Ranger*, who freely shared their current knowledge to supplement Paul Jones's dated information. Successful examples of irregular operations, both in wartime and peacetime missions, demonstrate the need for partners and show that their treatment as valuable teammates can be important to success.

In his biography of John Paul Jones, Samuel Eliot Morison explained that in the Whitehaven and St. Mary's raids "the damage done was inconsequential . . .

but the moral effect was stupendous."[109] And Jones himself pointed out the same in his report to the commissioners in Paris.[110] While the adjective "stupendous" may be strong, the analysis that the political and strategic results were more important than the tactical is accurate, as were Morison and Paul Jones's conclusions that the attacks had broader implications than the calling out of a fire brigade. "The pirate Paul Jones" captured the imagination of the British public, striking fear into communities along the coast and offering political capital to opponents of the government and the war. Because of its interaction with the population, the attack had much larger ramifications than taking enemy merchant shipping, which in many ways only appeared to the average Briton to affect the merchants and the insurers at Lloyd's. Attacks on British commerce impacted the Empire in fundamental ways and served as an important strategic element for the Americans. However, Paul Jones's direct attack on the population, a population that thought it was safe and well protected by the Royal Navy, seized their attention and created a political challenge for the leaders of an island nation, which also translated into a military problem. Clearly, however, the irregular missions alone were not sufficient for success. A total focus on the population in irregular warfare invites failure. If *Drake* had defeated *Ranger*, or the Americans had not escaped the other Royal Navy ships, the British people and subsequent historians would have seen the mission as a dismal failure. Instead of fearing the possibility of American attacks, the people of Great Britain would have felt secure that the Royal Navy would protect the coastal communities.

The raiding operations conducted by John Paul Jones during the American Revolution offer a starting point for the study of the American navy's conduct of naval irregular warfare in the Age of Sail. Despite Theodore Roosevelt's use of Paul Jones's image to support his own ambition for a large and conventional blue-water force, Paul Jones's missions off Nova Scotia and in the Irish Sea, and his planned operations on the Gulf Coast and off Africa, demonstrate wartime raiding had a place not only in his personal history but also the history of the early American navy. The Americans fought the revolution from a position of military weakness and relative disorganization. Despite this, they did not abandon naval irregular warfare once the United States Navy was officially formed and began to take a more organized, if still weak when compared to older kingdoms and states, role on the world's oceans. In fact, in the U.S. Navy's first two wars, with the forces of Revolutionary France and the Barbary Corsairs, naval irregular warfare again played a central role.

CHAPTER 2

Wars Done by Halves

Quasi-War Operations

Following the American Revolution, the United States sold off the ships of the Continental Navy. In the case of *America*, John Paul Jones's last command and the only revolutionary-era ship of the line completed in an American shipyard, the ship was given to France and sailed to join Louis XVI's navy in June 1783. When it came to naval affairs, political debates over the need for and ability to fund a navy, rather than the actual construction of a force, dominated the Articles of Confederation period of early American history. Despite the inclusion of Congressional responsibility to "provide and maintain a Navy" in the new Constitution, the decade following ratification saw continued political wrangling and inaction. Attacks on Americans in faraway seas finally forced the new government into an interest in naval affairs.[1]

The 1793 truce between the Kingdom of Portugal and the dey of Algiers put American maritime interests in the Mediterranean in danger. Britain wanted to coordinate with the Portuguese navy in their war against revolutionary France but Portugal's conflict with Algiers kept their ships busy patrolling the western Mediterranean and Strait of Gibraltar. British diplomats helped negotiate the 1793 agreement so that the Portuguese would be free to help fight the French. This, arrangement, however, had the added benefit of punishing the Americans. Following independence from Great Britain, American ships had lost the protection of the Royal Navy and the British diplomatic relationships with the Barbary sultanates of the North African coast. In their conflict with Algiers, Portugal's naval patrols had kept pressure on the corsairs, which offered the second-order effect of protecting American merchants and sailors. However, with the twelve-month truce in place, a squadron of Algerian corsairs made up of four small frigates, a brig, and three xebecs cleared Gibraltar and fell upon American ships in the eastern Atlantic and on the Iberian coast. By the time warning of the truce spread in the United States, ten American ships and their crews had been captured, conveyed to Algiers, and held for ransom.[2]

Congress passed the Naval Act of 1794 a few months later, authorizing the construction of six frigates to address the attacks in the Mediterranean and eastern Atlantic. However, the political bloc that opposed the development of a navy included a caveat in the bill that funding and construction would cease if diplomats negotiated a peace with Algiers. They authorized David Humphreys, the U.S. minister to Portugal, to negotiate with the Dey of Algiers. In September 1795, he signed a Treaty of Peace and Amity, which agreed to a tributary payment in exchange for protection of American interests against corsairs. Anti-naval politicians moved to halt the construction of warships that had begun in shipyards at Portsmouth, Boston, New York, Philadelphia, Baltimore, and Norfolk.[3]

Despite American interest in avoiding foreign entanglements in order to focus on building the new nation, as President George Washington soon illustrated in his farewell address, the outside world refused to be ignored. In the period between the authorization for naval armaments and the treaty with Algiers, the French Revolution appeared to pull the relatively new United States back into the long conflict between Britain and France, in which the American Revolution had played an earlier part. The signing of Jay's Treaty with Great Britain in late 1794, and American declarations of neutrality, angered the revolutionary government in Paris. French revolutionary leaders, and many Americans including Thomas Jefferson, saw the events in Paris as deeply connected to the ideals of the American Revolution. In that effort, the Americans had found an ally in France. That alliance had been formalized and made explicit via treaty. The revolutionary French believed the formal alliance was still in effect and would bring America to their aid, but President Washington disagreed and declared neutrality in the war between Britain and France. Many in Paris saw Jay's Treaty, despite its explicitly economic nature, as proof that America had not only become neutral but had sided with Britain.

American merchantmen carried goods to both sides in the conflict under their neutral flag, but the French reversed their previous policy of treating American ships as allies. Secretary of State Thomas Pickering reported to Congress that in 1795 French privateers and cruisers captured 316 American ships. The French colonial leaders in the West Indies had begun to commission privateers and pursued an aggressive course with little specific control from the Directory in Paris. While the Algerian danger had ebbed, French attacks flooded the West Indies and dramatically increased the threat to American ships and sailors.[4]

Washington attempted to address the conflict between the caveat language of the Naval Act of 1794 and the rising challenge from revolutionary France. In a March 1796 message to Congress he explained he was ordering a halt in naval construction, as the law required, but he believed stopping would have a negative economic effect in the communities around the shipyards. Combined with the French threat, his economic argument was enough to move the political leadership. Congress acted in March of 1796 with a compromise: authorizing the president to complete construction on three of the frigates. The 36-gun *Constellation* and 44-gun *United States* and *Constitution* continued to be built in Baltimore, Philadelphia, and Boston, while construction on *President*, *Congress*, and *Chesapeake* came to a halt.[5]

By 1797 John Adams had assumed the presidency and relations with France had further deteriorated. The French Directory, as the executive of the committee-based government of revolutionary France was known, refused to receive Charles Cotesworth Pickering as the new American minister to the French Republic. The French then made a formal declaration rejecting the principle that free ships made free goods, as recognized in the Franco-American treaty of 1778. In so doing they removed all restrictions on their privateers with regards to American trade. Adams sent a negotiating team to Paris to meet with the Directory's new minister of foreign relations, Charles Maurice de Talleyrand-Perigord, but talks fell apart in what became known as the "XYZ Affair." Congress began putting the United States on a footing for a maritime war.[6]

In April 1798 the Senate and House passed "An Act to Provide an Additional Armament for the further protection of the trade of the United States; and for other purposes," which authorized President Adams to buy or build a dozen small combatants of not more than twenty-two guns each, and officers and crew to man the vessels. Recognizing the speed at which relations with the French were deteriorating, the act even allowed the president to appoint the officers without Senate confirmation, in the event that they were needed during a congressional recess. Three days later Congress passed an act officially establishing the Department of the Navy, the administration of the sea service having previously been the responsibility of the Department of War. The formation of the Navy Department gave the government the organizational structure it needed to administer and fight a naval war. President Adams appointed the Georgetown merchant, and former Continental Army quartermaster, Benjamin Stoddert to the position of secretary of the navy.[7]

Congress continued the preparations for conflict. It passed legislation to increase American coastal defense capabilities, placed the cutters of the Revenue Service under navy command, and authorized letters of marque and privateering commissions.[8] The commissions for privateers authorized armed vessels, which sailed for the purpose of hunting enemy shipping, while letters of marque authorized merchantmen engaged in regular trade to arm themselves and attack targets of opportunity during their regular voyage. The most significant legislation passed on 28 May when armed conflict was authorized. In what might be seen as the beginning of an American theme of Congressional wavering over use of its war powers, they did not declare war on France. Instead Congress simply authorized

U.S. warships to attack and capture armed vessels "sailing under the authority or pretence of authority from the Republic of France" who might be "hovering on the coasts of the United States for the purpose of committing depredations on the vessels belonging to citizens."[9] They expanded the conflict in early July to "any armed French vessel, which shall be found within the jurisdictional limits of the United States or elsewhere on the high seas." Finally, in mid-July, Congress authorized the funds to complete the construction and launch of the frigates *President*, *Congress*, and *Chesapeake*, which President Washington had halted in accordance with the law in 1796.[10]

As *United States*, *Constellation*, and *Constitution* completed their fitting out, having launched during late 1797 and early 1798, the first U.S. naval vessel put to sea. The 26-gun, ship-rigged *Ganges*, a 504-ton merchant ship with a reputation for speed, was purchased on 3 May 1798 under the president's new authority and as one of the Navy Department's first official acts. The ship's merchant captain was Richard Dale, who had served as John Paul Jones's first lieutenant on board *Bonhomme Richard* during the Revolution. He received a captain's commission in the U.S. Navy and retained his command, armed his ship, and put to sea in less than a month. The United States Navy was established and ready for combat operations.[11]

However, the conflict with France was a confused one. Neither nation had declared war on the other, though both had abrogated their previous treaties. When reports reached France in the summer of 1798 that the Americans had decided to break off negotiations and open hostilities, the Directory realized they had made a mistake. However, a number of issues conspired to keep the two republics in a state of open conflict for almost three years. First was the slow speed of transatlantic communication, which caused attempts at negotiation to fall behind the events of the war. Second, the fact that the French Republic itself was still unstable and continued to have regular changes in leadership caused a pendulum effect when it came to efforts to calm the conflict. Changes in the makeup of the Directory's leadership swung the attitudes in Paris from concilia-tory to bellicose. And third, the Caribbean basin at the close of the eighteenth century was a dangerous place, with numerous competing powers, revolutionary efforts (such as in Haiti), and violent stateless and criminal groups, including pirates and smugglers.[12] The conflict thrust American naval officers and their ships into a hybrid conflict where war, politics, naval diplomacy, revolutionary fervor, and race-based fears of slave revolt complicated the maritime context in which they operated. It was a difficult and irregular conflict. Little wonder that Capt. Thomas Truxton related to Secretary Stoddert, after his defeat of the French frigate *L'Insurgente* in early 1799, "[T]he French Captain tells me, I have caused a war with France, if so I am glad of it, for I detest things being done by halves."[13]

John Adams and American leaders envisioned the conflict between the U.S. Navy and France as one of history's few pure examples of guerre de course. It was a limited conflict from both sides, neither the Americans nor French wanting it to expand beyond maritime trade war. They also envisioned it as a blue-water conflict. The May 1798 "Act to more effectually protect the Commerce and Coasts of the United States" authorized American warships to attack any armed French vessel, but limited those operations to "the high seas," which were defined as beyond one maritime league (three miles) from the coast.[14] The orders that Secretary Stoddert issued his captains included these rules of engagement. Operationally this meant that early in the war American captains envisioned two kinds of missions: patrolling open water for French cruisers and privateers, and convoying American merchantmen.[15]

At first, the United States prioritized sending larger cruisers to sea. The first three frigates were augmented by heavily armed sloops or corvettes like *Ganges* and the 20-gun *Delaware*, another converted merchantman commanded by Capt. Stephen Decatur Sr.[16] In the early months of the conflict American vessels began by patrolling the eastern seaboard of the United States. But at the end of July 1798 Secretary Stoddert wrote to President Adams with a plan to begin deploying American squadrons into the West Indies to catch the French cruisers near their colonial bases, particularly Guadeloupe. The ships would deploy together and fall under the command of a single commodore, but generally operated independently once in their area of operations. Stoddert began efforts to construct what would become the U.S. Navy's first rotational deployments in an effort to keep warships on station and in constant contact with the enemy, which he reported as only having a small force of warships in the region.[17] Over the late summer and autumn months American operations moved south and by December of the first year of the war President Adams had approved shifting almost the entire conflict into the West Indies, with squadrons deployed to four areas of operations.[18]

Over the course of the next eighteen months the U.S. Navy followed the pattern laid out by Stoddert and Adams. Ships deployed to geographically concentrated squadrons in the Caribbean, eventually supported by local naval agents to help supply them. A few battles occurred with privateers, the Americans made numerous captures, and there were a handful of engagements between American and French warships. These included Truxton and *Constellation*'s much-heralded defeat of the frigate *L'Insurgente*. Their attack on the other French frigate in the region, *La Vengeance*—a partial victory since the French escaped with severe damage by running to a neutral port—followed the next year. However, after operating in the theater American officers realized that while heavy American frigates were good for engaging the warships of the French navy, they were less effective in the archipelagic shallows of the Caribbean, which demanded numerous small combatants to chase down and engage the privateers that swarmed the area.[19]

Capt. Silas Talbot arrived at Cape François station in October 1799 and assumed command as commodore of the squadron from Capt. George Little. Truxton's defeat of *L'Insurgente* earlier in the year had significantly diminished the French naval presence. In January 1800, *Constellation*'s engagement with, and near victory over, *La Vengeance* would nearly end the ability of the French navy to engage in the conflict. Talbot's squadron would soon include not only his flagship *Constitution*, but also *Boston, General Greene, Herald, Augusta, Experiment, Richmond, Patapsco, Trumbull,* and *Norfolk*.

With the largest of the West Indies squadrons under his command, Talbot also had the largest number of competing priorities. With the French naval presence in the Caribbean nearly eliminated, the greatest threat from warships would come in the form of reinforcement from Europe. The trade winds and traditional sailing routes would bring any French warships to a first landfall at Hispaniola, likely at Cape François. As a result, Talbot could not let his heavier ships stray too far from this central rendezvous point. Yet privateers sailed from the blue water east of the Bahamas clear across to the coast of Cuba, all of which were within the squadron's cruising grounds. To complicate matters, Spanish privateers from Cuba had joined the French. Talbot and his crews also had to deal with the civil war that was simmering within Haiti. The forces of André Rigaud controlled the southern shores of the former colony, including the long Peninsula stretching away from Port-au-Prince and the Île de Gonâve. In the waters surrounding their territory pirate nests had developed, with barges and small boats swarming and raiding against any vessel that approached close enough. President Adams and Secretary of State Pickering had also adjusted American policy to support Toussaint Louverture in his effort to unify Haiti. Add to these complicated waters British interests that sometimes aligned and sometimes conflicted with American policy, and American merchants who disregarded their nation's laws to trade with anyone and everyone who was willing to pay, and historian Michael Palmer's suggestion that Talbot faced "difficult operational problems" seems a bit of an understatement.[20]

———————

On New Year's Day, 1800, the American schooner *Experiment*, under the command of Lt. William Maley, found itself becalmed in the Bay of Léogâne on the southwest coast of Haiti. The "bay" is not much more than a bight on the northern side of Haiti's long southwestern peninsula, which extends away from Port-au-Prince. *Experiment* was carrying U.S. Consul General Edward Stevens to meet with Haitian revolutionary leader Toussaint Louverture and had taken four merchantmen under convoy while sailing from Cape François to Port-au-Prince. The convoy had come around the western side of the Île de la Gonâve, when their wind died overnight within sight of land. As the sun rose the crews on board *Experiment* and the convoyed brig *Daniel & Mary*, and schooners *Mary*, *Sea Flower*, and *Washington* watched as men loaded into ten barges and set out toward them from shore.[21]

Maley had taken command of *Experiment* on her commissioning in Baltimore just two months prior. She was a 135-ton schooner built at the same time and to roughly the same dimensions of her sister ship *Enterprise*. Built for the shallow-water operations of the conflict with France, and under the command of aggressive junior officers, both ships saw great operational success in the Quasi-War.[22] Maley himself, while aggressive, also developed a reputation for being fast and loose with the rules of engagement. This was on top of being a harsh leader who was disliked among his peers and seniors. Secretary Stoddert, who had come to the conclusion that the young officer was a "very ignorant, illiterate man," discharged him from the service after his first cruise in command of *Experiment*. Despite his later failings, the engagement in the Bay of Léogâne proved to be what Michael Palmer characterized as "the only bright spot" in his short career.[23]

The waters where the convoy had becalmed were a known haven for pirates and violent actors in the Haitian civil war. Forces under André Rigaud, one of Toussaint Louverture's main rivals, controlled the southern end of Haiti except for Port-au-Prince. Many of his supporters practiced piracy using small boats in swarm attacks on merchantmen in the nearby waters. According to Capt. Silas Talbot, who took command of the American squadron off Haiti in October, it was the most dangerous part of the coast for American ships. When the convoy spotted the approaching barges, they were well aware of what was coming.[24]

As the boats approached from the shore the crews and passengers counted upward of five hundred attackers, with between thirty and sixty men in each of the barges. Armed with muskets, sabres, and boarding pikes, some of the larger boats had small guns on board or swivels mounted on their bows. Maley recognized that his position at the rear of the convoy was a poor place to protect his charges. He deployed his sweeps and his men slowly rowed *Experiment* into the center of the line. The crews had time to prepare since it took the attackers almost an hour to bring themselves in range. But Maley instructed his men to keep their gun ports closed and move slowly, in an effort to appear just like the merchant ships around them and draw the enemy in range. On board the brig *Daniel & Mary* and the schooner *Sea Flower*, which were each armed with three small guns, the captains prepared their decks for action.[25]

The barges rowed hard toward all five of the American ships. When they came within musket range the attackers began firing both their small arms and guns at the Americans. Maley ensured his gun crews were ready and after he assigned their targets he ordered *Experiment*'s broadside of six 6-pounder guns to open fire with grapeshot. The marines rose above the bulwarks and began firing with their muskets. *Daniel & Mary* and *Sea Flower* followed *Experiment*'s lead and commenced fire as well. Consul Stevens reported "our grape shot and small arms made dreadful havoc among them."[26] Surprised by the hardy defense, and realizing they were dealing with a small warship ready for combat, the barges pulled away. They came to rest just outside of cannon range and floated as a group; the leaders seeming to consult about what to do.[27]

A second small group of boats joined the attackers from the shore and appeared to put fresh men into the barges and take out the dead and wounded. After an hour and a half of regrouping, the barges hoisted their masts and sails and ran up flags. One group ran up the tricolor of the French Republic, likely indicating their identification with Rigaud, who considered himself aligned with the Directory in Paris, while others hoisted the blood-red banner associated with piracy in the Caribbean. The U.S. naval officers considered themselves at war with both flags. The pirates divided into three divisions and organized their second assault around subduing *Experiment*. Maley again had a short amount of time to prepare his crew as the boats rowed toward them. He placed his marines and some sailors with small arms on the bow and stern, since he had no chaser guns in either location, and he kept a handful of sailors on the sweeps in order to pivot his broadside in the calm air. Once the pirates approached within half a musket shot *Experiment* opened fire.[28]

The second assault spanned two hours. The pirates rushed *Experiment*'s sides repeatedly, each time driven back by her gun crews and marines. In the battle the Americans sank two of the attacking barges outright, and eventually turned away the assault entirely. However, the pirates did have some success. Around 11 o'clock, as some of them began to retire and it looked like the effort against *Experiment* would fail again, three boats split off from the main attack and rowed for the *Daniel & Mary* and *Mary*. The crew on board *Daniel & Mary* defended themselves successfully, with some supporting fire from *Experiment*'s gunners. However, the schooner *Mary* had no guns and the crew had no way to defend themselves besides personal weapons. With an overwhelming number of attackers coming for them, *Mary*'s crew hid in the ship's hold or jumped overboard as the barges approached. Captain Chipman of *Mary* was wounded by a musket shot to the arm and was the only American left on deck as the pirates swarmed over the sides and "inhumanely murdered" him, as two different witnesses described.[29]

On board *Experiment* they watched Chipman fall under the attacker's knives. Seeing no other Americans on deck, Maley directed his gunners to rake *Mary* with grapeshot as the pirates scoured the vessel. The fire, "proving too warm for the enemy," drove them off after they made off with Chipman's clothes and a small chest holding the ship's specie.[30] *Mary's* crew came out of hiding and cast their captain's body overboard, then took back control of the ship. The pirates retired to the beach and began to regroup yet again.[31]

The afternoon stretched on and the American ships remained becalmed. The merchantmen had no sweeps to move themselves and the tide began to carry *Washington* and *Daniel & Mary* away from the other three vessels. The pirates remained in view near the shore, hunched over their oars and watching the American ships as they separated. Captain Farley had expended all his ammunition on board *Daniel & Mary* and was as defenseless as Captain Taylor on board *Washington*. Both had witnessed the murder of Chipman from their decks and had a vivid image of what was in store for them. The ships drifted farther from *Experiment* and the barges began shifting in their direction. The two merchant captains elected to abandon ship. They placed their crews and passengers in their boats and crossed over to *Experiment*, where they were welcomed by the sailors and marines. Maley attempted to use his sweeps to move toward the abandoned vessels to continue to protect them but realized doing so would only take him farther from *Mary* and *Sea Flower*, both still fully manned. The pirate barges read the situation and rushed for the abandoned ships, towing them off toward their base. Finally, the breeze came up as evening approached and the remaining three ships sailed for Port-au-Prince.[32]

The only member of *Experiment*'s crew wounded in the engagement was Maley's first lieutenant, a young midshipman named David Porter. A musket ball grazed his arm. Porter went on to play an important role in the U.S. Navy in the Age of Sail and was a leading figure in the execution of naval irregular warfare, as will be seen in more detail in later chapters.[33]

The engagement between *Experiment* and the pirates aligned with Rigaud was the first combat during Capt. Silas Talbot's command as commodore of the Cape François squadron in 1799 and 1800. During this period, and the closing year of the war, operations off the coast of Hispaniola centered on naval irregular warfare operations. Counterpiracy and antiprivateer patrols continued in the west and southwest of the island following the attack on *Experiment* and her convoy. Yet Talbot knew that there was more to be done beyond convoy and blue-water patrols. Because he needed to be prepared for potential French warships to arrive from Europe, Talbot continued to maintain a concentration of his larger ships near Cape François, which was the likely first landing of a French crossing. But he also knew that the small and shallow-draft privateers would work the shallows around both the Spanish-controlled eastern end of the island and around the contested Haitian half. While the congressional authorization to attack French ships on the high seas remained in effect, Talbot elected to interpret his orders as expansively as possible.

The first week in May, *Constitution* drifted away from Cape François and patrolled off the Spanish northeastern coast. Talbot had received word that a well-known privateer was in port at Puerto Plata and set off to investigate. Knowing the *Constitution* was too large and with too deep a draft to patrol the shoreline closely, Talbot took the captured sloop *Amphitheatre* into American service as a tender for the frigate. He placed the recently promoted Lt. David Porter in command but reminded him that as the skipper of a tender, as opposed to an independent warship, he and his crew were still attached to *Constitution*. He ordered him off on patrol with orders that suspicious vessels "may be stop'd, and examined strictly, and brought to the Constitution if she is near to be found."[34]

Talbot left Porter and *Amphitheatre* patrolling off Monticristi on the northern shore and proceeded east toward Puerto Plata. The commodore was searching for a notorious privateer named *Sandwich*, a former British packet that French privateers had captured and armed with fourteen guns, at Puerto Plata. The *Constitution* arrived off the Spanish harbor late on 30 April and anchored out. The French privateer *Ester*, anchored in the harbor, spotted "a full rigged vessel frigate" in the morning along with the rest of the Spanish port and watched her closely. *Constitution* sailed easily off the port, counting the ships at anchor, which included the privateers *Ester* and *Sandwich*. Talbot sent his cutter with two lieutenants and some of the men on an early-morning reconnaissance of the harbor. They reported the significant shoals and reefs on the route into the anchorage and described the layout, including the small Spanish fort with four heavy guns. That night, having watered and collected her supplies, *Ester* slipped out of the harbor, rowing with her sweeps likely because she was afraid her sail would be more easily spotted. She rowed through the night to escape.[35]

Talbot had served in the army during the War for Independence and, despite commanding a privateer in that war, his fellow officers knew that he was a relatively poor shiphandler. However, his experience in land warfare on America's coasts had prepared him well to think about combat in the littorals and irregular operations. Serving under General Washington, he had organized the use of water transport to move troops into defensive positions during the New York Campaign of 1776, and he led a fireship attack on the British 64-gun ship HMS *Asia* in September 1776. In October 1778, he fitted out the small sloop *Hawke* and trained a crew of boarders to break the close blockade of Rhode Island. They attacked the 8-gun Royal Navy schooner *Pigot* in the night in a cutting-out expedition, taking the ship without the loss of a single American life. He then planned an assault on the 50-gun HMS *Renown*, but was unable to carry it off before winter set in and his boat was icebound. Naval irregular warfare had been central to his service in the revolution.[36]

Secretary Stoddert's knowledge of Talbot's strengths and weaknesses resulted in the assignment of Lt. Isaac Hull as his first lieutenant. In contrast to Talbot, the navy's officers regarded Hull as one of the best shiphandlers and brightest young officers in the new fleet. He would prove himself not only in the Quasi-War but also in the Barbary War and the War of 1812, becoming a central figure in early American naval history. Talbot wanted to sail *Constitution* into Puerto Plata and take on the four heavy cannon in the Spanish fortress and *Sandwich's* broadside. But consulting with Hull after sending the cutter in for the reconnaissance, they both concluded the shallows and unmarked reefs at the approach to the harbor made a straightforward attack far too dangerous.[37]

Ester continued to patrol the seas east of Puerto Plata and fell in with the American merchant brig *Nymph*, capturing her easily. In order to plunder the cargo, the two small vessels sailed into a bay on the northeastern tip of the island and met with a local barge. *Constitution* patrolled west from Puerto Plata, the opposite direction from *Ester*, and joined up with *Amphitheatre* and *Boston*. When the Americans returned off the harbor on the tenth, the lookouts counted the masts and reported that a ship had slipped out. Talbot ordered his vessels in pursuit to the east, likely believing they would have spotted *Ester* if she had sailed west. On 8 May *Constitution's* lookouts saw a strange sail in the bay at the northeast end of Hispaniola. Porter and *Amphitheatre* headed into the shallows to investigate.[38]

After sailing into the bay Porter came rushing back within hailing distance of the flagship. His crew had spotted *Ester*, *Nymph*, and the barge anchored in the bay. Talbot ordered a detachment of marines transferred into *Amphitheatre* and then *Constitution* manned four boats to "bring them to action, which they commenced with a running fight." As the tender and boats fought their way into the bay the combat lasted about thirty minutes before *Ester's* crew ran her in toward shore and attempted to ground the ship. Two dozen of the crew of forty-three privateers jumped overboard and scrambled up the beach, escaping into the bush. *Amphitheatre* tried to chase the barge as it raised the sails on its lugger rig, but navigating the larger American vessel around the reefs and shoals in the bay slowed their pursuit. Porter quickly gave up the chase and, as evening approached, he and his small expedition of boats pulled *Ester* off the sand bottom. *Amphitheatre* and the boats towed the privateer and *Nymph* back out to *Constitution* with three wounded Americans, three dead privateers, and sixteen captives.[39]

Despite the captures, Talbot was not ready to give up on the escaping barge. He re-manned and rearmed two of the boats and placed them under the command of Lt. Isaac Collins. At 1930, the small party rowed off into the night in pursuit of the three-masted barge that had escaped over the reef and shoals. *Constitution* took *Ester* in tow and placed a small prize crew on board *Nymph*. The ships set off at an easy sail into the darkness, paralleling the course to the south where they thought Collins's boats were going.[40]

Constitution sailed through the night and tacked off and on throughout the morning, waiting for Collins's boats to return. After lunch, they spotted the flagship's large cutter, which had accompanied Collins, but when it came alongside the midshipman in command related that he had lost Collins and the other boat in the night and had not seen them, or the escaped barge, since. Talbot ordered Porter to lead *Amphitheatre* and *Nymph* inshore to patrol in search of the barge for the rest of the day and night, and to return to *Constitution* the following day. Before the two small tenders returned from their patrol the following morning, *Constitution* fell in company with the 18-gun *Herald* from the squadron, which had been working the southern coast. Lt. Charles Russell, *Herald*'s commander, told Talbot that he had seen and spoken with Lieutenant Collins, whose boat had captured a small sloop. The two vessels had spent most of the night together, so the wayward raiders could not be far away. Two hours later, Collins came alongside in his captured sloop, towing the flagship's barge.[41]

The captured sloop was a 58-ton American trader named *Sally*, which had recently sailed from Puerto Plata and was scheduled to return there before sailing for its home port of Providence, Rhode Island. Collins and his raiders had discovered the sloop anchored in shallow water of the eastern shore of the island and suspected her of either being a privateer or a smuggler. They fell upon her in the dark as the crew slept, quickly and easily taking control of the ship. Collins listened to Thomas Sandford, the master of *Sally*, as he explained that the ship was American and that the naval officers should release him. However, seeing that U.S. neutrality laws forbade American merchants from trading with the island of Hispaniola outside of Toussaint Louverture's ports, and two-thirds of Sandford's cargo was already on shore, Collins judged the ship to be a smuggler and took her as a prize.[42]

On board *Constitution* the next morning, Sandford confirmed to Talbot that *Sandwich* was in Puerto Plata, protected under the guns of the Spanish fortress. Talbot and Hull had been concerned with the deep draft of *Constitution* in trying to cut out the privateer. They knew the locals were suspicious and since the ships anchored in the harbor had already seen their cutters during the reconnaissance the week before, mariners in the port might recognize them. But the capture of *Sally*, already scheduled to return to Puerto Plata, offered an opening they had not expected. The presence of the small Rhode Island sloop may have seemed auspicious to Talbot, since his privateer command *Argo* in 1789 had originally been a Rhode Island trader re-christened from her original name: *Sally.*[43]

Sandford's ship was small enough to make it into the harbor easily, and she was familiar to the watermen of the region. Because of her scheduled return, there would be no suspicions of her. Talbot confirmed Collins's decision that the small ship was a prize and ordered Hull to take command of her on an expedition into the harbor. Sandford agreed to go along as a pilot; he was either strong-armed by the officers or potentially hoping that if he helped, Talbot might return his ship to him. The commodore assigned Hull a force of ninety sailors and marines to sail *Sally* into Puerto Plata and cut out the French privateer. The crews emptied *Sally* of what remained of her cargo and stores. They redistributed the contents of the hold between *Nymph* and *Ester*, which Talbot sent off with a prize crew to the New York Admiralty Court. *Herald* left the small group of ships as *Ester* sailed north, and *Constitution* and her consorts remained well away from Puerto Plata in order to maintain surprise. Talbot wrote his orders to Hull for the expedition to "bring her out to sea, if practicable; otherwise to burn and destroy her in port." Hull was assigned *Constitution*'s two marine corps officers, Capt. Daniel Carmick and Lt. William Amory, to help lead the expedition. They loaded *Sally* with small arms and cutlasses and took the sailors and marines on board. Captain Carmick later recalled feeling like Achilles and his Greek warriors, the sailors and marines hidden in their own Trojan horse. Hull and six sailors remained on deck to work the ship, with Thomas Sandford standing on the quarterdeck for the world to see. The remaining force of sailors and marines hid in the hold. They set off for the Spanish harbor on 12 May.[44]

As evening approached, *Sally* was on course for the Spanish harbor, looking for all the world like a merchant sloop, when the Americans came across a British warship patrolling near Puerto Plata. The warship fired a gun in warning and sent a boarding party toward what they thought was a coastal trader. Hull welcomed on board the British lieutenant leading the boarding party. He quickly realized the purpose of the vessel, "much surprised on getting onboard the sloop to find the hatch ways filled with American officers and seamen." The British officer related that his ship had been watching *Sandwich* in the harbor and considering their own cutting out. It looked like they were too late for the prize. The lieutenant wished his American counterparts luck and rowed back to report to his captain.

Sally made short progress with little wind through the night. But a sea breeze sprang up in the morning and the raiders approached the harbor just before noon. Hull, with Sandford's advice as pilot, guided the disguised raider past the reef and into the harbor. *Sandwich* lay at anchor and with all her guns hauled to the starboard side to bear on the channel. Riding the noontime sea breeze, the Americans maneuvered through the sparsely populated harbor and brought *Sally* alongside the privateer's starboard bow.[45]

On Hull's signal, the men poured out of the hold and across the bulwarks. Captain Carmick led the raiders over *Sandwich*'s rail, cutlasses in hand. He later wrote: "[T]he men went on board like devils, and it was as much as the first Lieutenant and myself could do to prevent blood being spilt." The men fired six shots as they boarded, and the attack was so quick and overwhelming that the French privateer crew scrambled to hide in their ship's near-empty hold rather than fight. The captain of the privateer realized the hopelessness of his situation and, hat in hand, surrendered to Lieutenant Hull.[46]

Hull took stock of the captured vessel and confirmed that it was in the middle of refit. The French crew had stripped the ship, only its lower masts were standing, and the crew had coiled the rigging and sails and stowed them below. The Americans needed time to re-rig the ship in order to escape, but the Spanish fortress overlooked the anchorage with a constant threat from its guns. The officers ordered the prisoners bound and placed under the watch of several sailors, and Hull dispatched Captain Carmick and Lieutenant Amory with the marines to take the fortress.[47]

The marines lowered *Sandwich*'s boat and embarked their small force, which rowed for the shore at the base of the fortress. As they approached the rocks, the marines climbed into the neck-deep water, holding their muskets and cartridge bags above their heads, and waded ashore. They rapidly moved into the fortress before the Spanish realized what was happening. With surprise on his side, Carmick and his marines took possession of the fortress and spiked the guns. The marines then returned to the deck of *Sandwich* and took up small-arms positions and manned some of the guns, which had been hauled over the face shoreward, to protect *Sandwich* and *Sally* while the sailors worked to rig the ship. By nightfall, the crew had raised the topmasts, rigged the ship, and bent on the sails.[48]

When the rigging was completed, the Americans faced a new challenge: There was no wind. The sailors manned the ship's guns, and the marines took up positions as sharpshooters while the Spanish garrison mustered on shore to face them. Their cannon rendered useless by the marines' raid ashore, the Spaniards had no way to challenge the Americans other than their muskets. As night set in, the local commander sent several flags of truce by boat to determine the raiders' intentions. Hull and Carmick simply replied that they were under orders of their commodore and they would remove *Sandwich* from the harbor.

When morning approached, a land breeze sprang up and the Americans set their sails and weighed anchor. They sailed clear and joined *Constitution*, without interference from the Spanish garrison.[49]

Talbot reported the operation to Secretary Stoddert and praised his officers. Likely remembering the parallels with his own exploits against *Pigot* many years before, he wrote, "[N]o enterprise of the same moment was ever better executed. The operation was completed without a single casualty." The copper-bottomed *Sandwich*, which had been preying on American merchantmen for three years, was taken out of action and now belonged to the Americans. However, while it was a tactical success because Hull and Carmick brought out the ship without losing a man and perhaps an operational success because *Sandwich* was no longer a threat to American shipping, at a strategic level results were much less clear. Talbot had sent an armed vessel into a Spanish port and his men had even assaulted a Spanish fort. The prize was returned to the United States for adjudication in Admiralty Court. There, things were not nearly as easy for Talbot to finesse as the tactical thinking on the scene.[50]

In early July, a formal protest from the Spanish government reached Washington, D.C., over the violation of the neutrality of their harbor. By that time, *Sandwich* was in the custody of the New York Admiralty Court, and the disposition of the case was delayed as the State and Navy Departments responded to the Spanish. Capt. Alexander Murray relieved Talbot as the commodore of the squadron as *Constitution* rotated home in the early summer of 1800, and the new commander wrote to the secretary that, from his understanding of the state of affairs on Hispaniola, the challenge from the Spanish was probably in the right. Stoddert decided it would be better for the executive branch to agree with the Spanish and offer to return the ship as an olive branch, rather than let the courts decide the issue. The Navy Department ordered *Sandwich* returned to her owners, who sued for damages and eventually won in American courts. However, the expense of the damages was made up for by the judgment against *Nymph* and *Sally* as legitimate prizes. Despite this apparent rebuke, Secretary Stoddert assured Talbot that he understood the confused diplomatic issues of the theater meant that mistakes would happen, and that despite the apparent rebuke he assured him that it took nothing from the honor or glory of him or his crew for having carried out such a successful operation. The secretary was apparently willing to forgive some diplomatic and legal fudging of the rules of engagement so long as it came from well-meaning aggressive prosecution of irregular operations. In the end, by the time the courts returned *Sandwich* to her owners and she could cruise again, the war had ended with the successful negotiation of the Treaty of Mortefontaine.[51]

Despite Thomas Truxton's lament of wars done by halves to Secretary Stoddert in early 1799, the history of the first American naval war is a history full of irregular operations that played a role in the planning and execution of the conflict. In the West Indies, operations other than blue-water ship-on-ship or squadron-level engagements were a significant part of American efforts. Both the Quasi-War and the Barbary War that would follow were undeclared wars, where congressional action waffled between authorization of combat with funding for naval operations and a formal declaration of war. These kinds of limited conflicts have a long pedigree in world naval history, even making up a significant element in Sir Julian Corbett's early-twentieth-century work on naval strategy. They also demand serious attention to irregular operations.[52]

The naval irregular operations conducted in the Quasi-War have similarities with the foundations laid by John Paul Jones's operations with the Continental Navy two decades prior. The history of Paul Jones's operations at Whitehaven and on the coast of Britain demonstrated the importance of particular leadership traits, the interaction between conventional blue-water and irregular operations in the conduct of naval conflicts, and the need for local knowledge and intelligence or partnerships. The American experience against the French and against the factions in the Caribbean from 1798 to 1801 illustrates these elements as well.

Silas Talbot offers an example of senior leadership in irregular warfare. He balanced unique and challenging diplomatic elements of his orders with an aggressive mindset, and trust in his lieutenants. Talbot's personal experience in the War for Independence naturally inclined him toward understanding and embracing naval irregular warfare and raiding operations. His cutting-out of *Pigot* and the fireship attack on HMS *Asia*, as much as his leadership in privateering, gave him both an intimate understanding of the tactical and operational realities of irregular operations and the knowledge that he had to trust his men. Paul Jones's experience prior to his command in war had been limited to that of a mariner, but as later chapters will demonstrate, experience in naval irregular warfare became a central way to teach responsibility and leadership to junior officers in the Age of Sail. Midshipman Porter and Lieutenant Hull, who were central players in the operations described in this chapter, will appear again in the history of naval irregular warfare as the nineteenth century progresses.

In the West Indies conventional naval missions, like convoy escort and blue-water patrols, were important elements of the war. Without Truxton and *Constellation*'s victories over the French frigates *L'Insurgente* and *La Vengeance*, Talbot would not have had the freedom of action he needed for his irregular operations during his turn as commodore of the Cape François squadron. Likewise, other squadrons in the West Indies kept ships on patrol in the western Atlantic and blue-water approaches where the larger French privateers sailed. The experience of the U.S. Navy's first conflict reinforces the observation from Paul Jones's cruise on board *Ranger* that naval irregular warfare should be seen in balance with conventional operations.

Because conventional operations remained a part of balanced operational planning, there is a tendency to weight large multimission warships heavily in fleet design. But the commodores in the West Indies during the Quasi-War regularly found themselves needing another type of ship to support their operations: small combatants and gunboats. The frigates, which were the focus of shipbuilding and fleet design for the United States Navy, were large, well-constructed, and well-armed ships that were a match for any comparable ship. This was demonstrated by Truxton on board *Constellation* and was reinforced in the blue-water combat of the War of 1812. However, small combatants continued to make up too small a portion of the American fleet. Once Congress authorized combat in the Quasi-War, the Americans scrambled for small combatants, buying and converting merchantmen and drawing revenue cutters into the fleet. It still was not enough. The commodores regularly took the small craft they captured into American service to fill the gaps in their capabilities, from Truxton's capture of *L'Union* off Guadeloupe to Talbot's capture and use of *Amphitheatre* and *Sally* as tenders and shallow water ships. Captured prizes were the platform of choice in the irregular operations but also illustrate the importance of balance in fleet design. The small ships were used for the irregular missions but also had important roles in conventional operations patrolling and protecting convoys.

The dissolution of the Continental Navy and the founding of the United States Navy a decade later did not diminish the importance of naval irregular warfare and raiding operations in the early American naval experience. The first war fought by the new U.S. Navy continued to illustrate the centrality of irregular operations in naval affairs and built upon the principles seen in Paul Jones's experience in the Irish Sea and on the southern coast of Britain. It was mere months before the officers, who had been patrolling the shallows and raiding against pirates, smugglers, and privateers in the West Indies, would be at it again, this time more than four thousand miles away on the shores of Tripoli.

CHAPTER 3

INTREPID AND IRREGULAR
WARFARE ON THE BARBARY COAST

The Convention of Mortefontaine, completed in September of 1800 with the treaty ratified by the Senate in February 1801, could not have come soon enough for the defense of American interests. By the spring of 1801, the United States Congress was again debating how to reduce or even eliminate the U.S. Navy that they had built for the conflict with France. Yet it was only a matter of weeks before the security of American citizens demanded a naval response in another part of the world. On 1 May, world events again changed the thinking in Washington, D.C. The pasha of Tripoli sent his men into the courtyard at the American consul's residence and chopped down the flagpole bearing the American flag, declaring war on the United States of America.

At the turn of the nineteenth century, the Maghreb or northern coast of Africa—known to Europeans and Americans as the Barbary Coast—was made up of the independent sultanate of Morocco and the regencies of Algiers, Tunis, and Tripoli, which owed allegiance in one form or another to the Ottoman Empire. These tributaries, for the most part autonomous, were the home of a developed culture of piracy and slave trade that stretched as far back as the fall of the Roman Empire. During the eighteenth century, raiders from the Maghreb had captured over 150,000 European Christians who were then sold into slavery or held for ransom. By the time of the American Revolution, a well-established system of tribute existed by which the trading nations of Europe paid protection money to the North African rulers in return for the safety of their ships.[1]

In early 1801, Pasha Yusuf Karamanli of Tripoli knew that Algiers was earning a much higher tribute than he was, by the terms of the 1795 treaty the Americans had negotiated with the larger North African power. Karamanli had agreed to Tripoli's 1796 treaty with the United States, negotiated with Joel Barlow, only a few months after usurping his older brother to assume the throne. At the time, he had been in a weak position and needed to consolidate his power rather than start a conflict with the United States. However, five years later he was in a different position. Karamanli demanded an increased tribute of nearly $250,000. When the United States refused, he declared war.[2]

In response, President Thomas Jefferson sent a series of naval squadrons to the Mediterranean to protect American merchant shipping. The first deployment, which began in 1801 under Commo. Richard Dale, was marked by frustration and failure. The force, described as a "Peace Establishment" in the Congressional funding, operated under strict rules of engagement. Under their orders from Secretary of the Navy Robert Smith, American warships could intervene only when they directly witnessed an attack on an American merchantman by a corsair.[3] The primary success of the squadron came at Gibraltar, where it discovered Tripoli's naval commander anchored there with two vessels. Dale ordered Capt. Samuel Barron to lay off the harbor with the 36-gun frigate *Philadelphia*, bottling up the small Tripolitan force. The Tripolitans eventually gave up waiting for *Philadelphia* to leave, dismantled and sold their ships, and discharged their crews. Meanwhile, Dale dispersed his three other ships across the Mediterranean, where they conducted convoys and cruised singly for corsairs. In the end, because of both the small size of the force and the restrictive rules of engagement, they had little to show for their efforts.[4]

A second mission left Hampton Roads in 1802, under the command of Commo. Richard Morris. Although the president and Congress had expanded the rules of engagement, "Morris' squadron behaved more like a touring company than a naval force."[5] Morris brought his wife along for the trip and spent plenty of time ashore with family. William Eaton, the American consul at Tunis during 1802, asked in a letter, "What have they [the squadron's crews] done but dance and wench?" Morris's deployment was even less successful operationally than Dale's, despite having more ships and more aggressive rules of engagement. This inactivity, and reports of the squadron's behavior that reached Washington, resulted in Morris's relief and official censure. In 1803, the Navy Department dispatched Commo. Edward Preble to the Mediterranean with a third American squadron.[6]

There are a number of excellent histories of the First Barbary War. Glenn Tucker's *Dawn Like Thunder* and A. B. C. Whipple's *To the Shores of Tripoli* stand out as two of the best, with apparent archival research and historical analysis of the conflict but less than adequate notes. Some recent historical work about the conflict has suffered from strained efforts to connect the two-hundred-year-old events to the twenty-first century struggle against Islamic terrorism. It is not the purpose of this chapter to recount the entirety of the war, either militarily or diplomatically. Nor is it the intent to counter the strengths and weaknesses of the previous historiography. Instead, by focusing on the events surrounding the history of the small ketch *Intrepid*, née *Mastico*, the following pages examine the irregular warfare operations and raiding element of the conflict carried out by Commo. Edward Preble, Lt. Stephen Decatur, and the officers that became known as "Preble's Boys."[7]

The first great challenge that Edward Preble encountered when he reached the Mediterranean was the capture of the frigate *Philadelphia* by Tripolitan forces while he was busy negotiating with the Moroccans. In late October 1803, Capt. William Bainbridge was commanding the frigate on blockade duty off the Tripolitan coast. On the morning of the thirty-first, lookouts spotted a coastal blockade runner "very near the shore" running toward the harbor of Tripoli. *Philadelphia* gave chase and the vessel hoisted Tripolitan colors. Bainbridge ran in as close to shore as he felt comfortable, carefully checking his charts, which indicated forty-two feet of water beneath his keel. As the Tripolitan ship neared the entrance to the harbor *Philadelphia* was obliged to bear off the wind, allowing the small blockade runner to escape. Turning away from the chase, the frigate ran aground on unmarked rocks in twelve feet of water.[8]

The Americans were unable to refloat the ship. Bainbridge ordered his sailors to begin jettisoning cannon, stores, and ballast stone. As the crew worked to free the frigate, gunboats sailed from the fortress in Tripoli harbor and began to attack the grounded ship. The attacking boats fired high, aiming primarily for the rigging, which would help keep the frigate immobile, and Bainbridge recognized that if they had fired at the hull there was "no doubt but that they would have killed many."[9] Still unable to free the ship, the captain ordered the foremast cut away and cast overboard. It had no effect. According to a letter written by the frigate's wardroom, in defense of their captain's eventual surrender, "every exertion was made, and every expedient tried, to get her off and defend her."[10] Fearing for the safety of his crew as the enemy shot began to gain accuracy, and having cast his own guns overboard, Bainbridge surrendered his command to the Tripolitans. Local knowledge of the tides and currents allowed the enemy to float the ship shortly after the gunboats transported the prisoners, where the victorious Tripolitans paraded them through the streets. Eventually, local vessels towed *Philadelphia* into the harbor as a prize of war.[11]

The capture overshadowed Preble's successful negotiation with Morocco, where the sultan reaffirmed the terms of his previous treaty with the United States. Word of the attack reached him while laying off Malta in November and he realized he faced a difficult situation. The international prestige of the United States, and the Navy in particular, plummeted after the capture of the frigate. Following the ineffective efforts, or lack of effort, of the first two squadrons, American naval leadership appeared at best foolish, at worst incompetent. Preble could try and mount an invasion of the city of Tripoli to retake *Philadelphia* and free Bainbridge and the prisoners, or he could find another solution.[12]

Intelligence reports from inside the city indicated that *Philadelphia* lay deep in the harbor, "within pistol shot of the whole of the Tripolitan marine, mounting altogether upward of one hundred pieces of heavy cannon, and within the immediate protection of formidable land batteries, consisting of one hundred and fifteen pieces of heavy artillery." In the harbor a mixed fleet of nineteen gunboats, two schooners, two galleys, and a brig lay at anchor or moored to the quay, with over a thousand Tripolitan sailors on board. On top of the naval forces, there were the guns of the fortress, which estimates said were supported by 25,000 troops encamped around the city. Preble's squadron had only seven ships and eight gunboats. Even including the detachments of U.S. Marines on board the ships, he could only muster a total of 1,060 men. The commodore realized that there was little chance of success if he mounted an invasion of the pasha's city with the forces that were available to him.[13]

It was only a matter of weeks before the commodore began receiving letters from Bainbridge in Tripoli, which provided valuable intelligence. Under guard in the former home of the American consul, Bainbridge had devised a cipher and invisible ink in order to send innocuous-appearing letters to his commodore. The Danish consul, who had taken it upon himself to help the American prisoners, sent the correspondence out of the city. In one of his first messages Bainbridge reported the state of the harbor defenses and told Preble that "I think it very practicable with six or eight good boats well manned, and determined officers to destroy her [*Philadelphia*], and their cruisers, particularly if the thing was attempted without giving them much warning." He went on to suggest that "by chartering a Merchant vessel, and sending her into the harbour, with men secreted and steering directly on board the frigate, it might be effected without any or a trifling loss." Bainbridge's suggestion drew on the history of American maritime raids, and Preble was already considering similar options. Before receiving the coded letter with Bainbridge's suggestion, he had already promised Secretary of the Navy Robert Smith that he would "hazard much to destroy her." Like Talbot off Puerto Plata, Preble was likely recalling his service in the Revolution where, as a lieutenant on board the Massachusetts Navy ship *Winthrop*, he led a successful raid and boarding operation to capture the 14-gun British brig *Allegiance* from a harbor on the coast of Maine.[14]

A month after the capture of *Philadelphia* the imprisonment of her crew, the frigate *Constitution* and the schooner *Enterprise* sailed to assume the blockade duty outside of Tripoli. An aggressive young lieutenant named Stephen Decatur Jr. commanded the smaller ship, which had been built in Baltimore in 1799 for the Quasi-War. Decatur was a second-generation American naval officer; his father, Capt. Stephen Decatur Sr., had commanded the 20-gun *Delaware* in the Quasi-War and made the first capture of a French vessel in that war. He had also been the first commanding officer of *Philadelphia* and had commissioned the frigate in 1800. Decatur and his crew on board *Enterprise* had already experienced a successful deployment by the time winter approached in 1803. That fall they had captured several Tripolitan vessels and took part in the bombardment of enemy ports. On 23 December 1803, as the sun began to rise, the ship was nine miles east of Tripoli harbor and sailing in company with *Constitution*. The lookouts reported land in sight, and only minutes later a lookout spotted a sail on the horizon. Decatur signaled the discovery to *Constitution* and set off in pursuit. Commodore Preble on the flagship set a course to follow *Enterprise* as the smaller ship headed into the shallower water to intercept the unknown vessel.[15]

At 0830, the winds shifted to the southwest and the unidentified vessel changed course with the wind to remain ahead of *Enterprise*. The new course drove the pursuit toward the deeper water where *Constitution* sailed. Decatur and *Enterprise* caught the ketch after an hour and a half in chase, as they approached the frigate, and fired a warning shot, which caused the runner to heave to. The crew identified the unknown vessel as *Mastico*, sailing under a Turkish flag. The small 64-ton ketch was 60 feet long and had a 12-foot beam. The ship had been constructed in France in 1798 to serve as a bomb ketch during Napoleon's Egyptian expedition and was sold to merchants after the failure of the French campaign. It had been plying the North African coast ever since.[16]

Decatur put a boat in the water and sent a boarding party across to *Mastico* while *Constitution* remained within gunnery range. *Enterprise*, the closer of the two warships, covered the boarding party with marine sharpshooters and swivel guns. The party from *Enterprise* boarded the ship and discovered a mixed crew of Turks, Greeks, and Tripolitans. While it flew the Turkish flag the vessel's documents and logs indicated that it and the cargo were from Tripoli, and therefore it was likely in violation of the blockade. Despite his initial attempt to flee, the ship's master appeared compliant once he found himself under the guns of a 44-gun American frigate.

Commodore Preble ordered the vessel brought alongside *Constitution* and the Americans began a more detailed search of the ship. Despite the fact *Mastico*'s master claimed to be a merchantman, they discovered two cannon on deck, two more in the hold, an armory full of pistols and muskets, and Tripolitan military officers and soldiers on board. Preble found the cargo, a good portion of which was made up of slaves being transferred from Tripoli to Constantinople, suspicious and he believed that they violated the blockade but he did not have anyone capable of translating the vessel's Turkish documents to be sure. He also had the word of Pietro Corcillo, an Italian physician who had worked in the Karamanli court but had joined the Americans as a surgeon's mate after falling out with the pasha. Corcillo positively identified the two Tripolitan officers on board the ketch. In order to pursue the investigation further, Preble ordered Midn. Hethcote Reed to lead a prize crew on board *Mastico*. They sailed the captured ship to the American base at Syracuse on the island of Sicily.[17]

Once anchored in the harbor at Syracuse, the Americans confirmed the true identity of *Mastico*. Besides Corcillo, another man came forward who recognized the ketch. The second witness was Salvatore Catalano, a native of Palermo and an experienced Mediterranean merchant captain, who was at Syracuse when *Mastico* sailed into the harbor. He had recently returned from a voyage to Tripoli at the very end of October and was in the port when *Philadelphia* grounded. Anchored near his ship was *Mastico*, flying Turkish colors. When word of the grounding spread through the harbor the ship's captain, Mustapha Rais of Candia, lowered his Turkish flag and raised a Tripolitan one. He loaded a group of soldiers, then weighed anchor and sailed for the rocks where *Philadelphia* lay stranded. Rais and *Mastico* took part in the capture of the frigate and then the ketch helped carry the American prisoners back to the pasha's fortress. The captain personally helped escort Bainbridge and his men to see the pasha. Catalano's story was confirmed when a thorough stem-to-stern search of *Mastico* uncovered the officer's sword of Lt. David Porter, *Philadelphia*'s first lieutenant.[18]

Commodore Preble concluded, "[I]f a Tripolitan, he is a prize. If a Turk, a pirate." Either way, the vessel could be condemned. Preble collected the documents associated with the capture and his reports, then sent them back to the United States for adjudication in the court. The evidence included Catalano's testimony and the American officer's sword. Sure of the outcome of the court proceedings, Preble claimed *Mastico* as a prize of war. A survey of the ship placed its value at $1,800, and the commodore used the expected prize money to purchase the vessel for the government. Commodore Preble took her into the American service and renamed the ketch *Intrepid*.[19]

The Tripolitan officers and soldiers on board *Mastico*, as well as the slaves that were being transported, became prisoners of the Americans. Some of the slaves were intended as tribute gifts to the pasha general in Constantinople, and others were to be sold there for leading officers and merchants in Tripoli. The commodore's first inclination was to use the captured officers, soldiers, and slaves as bargaining chips to negotiate for the release of some of Philadelphia's sailors.[20]

Lieutenant Decatur was the first of Preble's subordinates to approach him with a specific plan to raid Tripoli harbor to take *Philadelphia*. He wanted to do it with his ship *Enterprise*. However, the ship was in poor material condition and in desperate need of refit. It was also probably too large for the job. Preble put him off but promised to keep Decatur's plan in mind. Bainbridge had suggested chartering a merchant vessel for a raid into the harbor, but as the investigation into the blockade runner closed and the Americans took possession of *Mastico*, all of the points of discussion and suggestion came together. The capture of the small coastal vessel was still recent and likely to be unknown in Tripoli's harbor, and it was small enough and blended easily with the local commercial traffic. Sending the American squadron in close enough to ensure *Philadelphia*'s destruction by bombardment would place its ships in danger from the massed enemy guns. The Americans also did not have enough force for a full assault on the city. With Decatur in command and men from *Enterprise* as crew on board *Mastico*, the captured ketch could slip into Tripoli's harbor, then board and recapture *Philadelphia*. Once Decatur was in control of *Philadelphia*, the lieutenant intended to sail it clear of the harbor and back into the service of the United States.[21]

Preble considered the new plan and realized that it just might work. On 31 January 1804, the commodore formally rechristened the captured ketch *Intrepid* and ordered Decatur to take command of an expedition against Tripoli harbor. He authorized Decatur to load stores for thirty days and draw a crew of seventy-five sailors and marines. Preble instructed Decatur to take only volunteers, for the mission would be dangerous. He sent him five midshipmen from *Constitution*, which included Midn. Thomas MacDonough of *Philadelphia*, who had been commanding a prize crew at the time of the frigate's capture, as well as the flagship's recently signed Italian pilot, Salvatore Catalano. The commodore was clear in the purpose of the mission: he wanted the *Philadelphia* destroyed—not recovered, as Decatur proposed. His formal orders to Decatur stated, "The destruction of the *Philadelphia* is an object of great importance"; they gave strict instructions that "after the ship is well on fire, [Decatur was to] point two of the eighteen-pounders, shotted, down the main hatch, and blow her bottom out." Attempting to sail the frigate clear of the harbor would be too great a risk to his men, no matter how gallantly Decatur led them.[22]

Preble also ordered the 16-gun brig *Syren*, commanded by Lt. Charles Stewart, to accompany Decatur and provide support as required. The navy had launched the small warship, built by the Pennsylvania shipwright Nathaniel Hutton, in Philadelphia on 6 August 1803. Stewart had outfitted and commissioned her into the U.S. Navy himself. Rigged as a brig, 94 feet long and 27 feet, 9 inches wide, *Syren* had a displacement of 240 tons. A crew of 120 sailors and marines manned the ship, which was armed with 24-pound carronades. Stewart and his ship sailed from Philadelphia on 27 August 1803 to join the squadron in the Mediterranean and had spent six months conducting wartime patrols under Preble.[23]

Decatur took two days to load the stores, weapons, and explosives before *Intrepid* set sail in the company of *Syren*. The 60-ton bomb ketch, designed as a coaster and generally unfit for long, blue-water voyages, had a difficult time crossing the 250 nautical miles from Syracuse to Tripoli. Also, the men on board *Intrepid* discovered, after they were under way, that many of the stores they had been issued were putrid and unusable. On 7 February, as they approached the North African coast, a gale struck the two American ships; *Intrepid*'s small size and poor construction nearly doomed the expedition. They survived the storm and the poor provisions, but word spread among the crew that the enemy may have discovered the force. The confidence of the men was severely tested. Lewis Heermann, a navy surgeon asked by Decatur to join the mission, wrote later that these challenges "laid the foundation of apprehensions of eventual failure."[24]

On 16 February, under a noonday sun, *Intrepid* approached within sight of Tripoli harbor. The weather had improved following the gale, but the horizon did not look promising and the crew suspected a second storm was coming. Decatur called a council of his officers to discuss their situation: dwindling stores, poor weather approaching, and a crew that was beginning to lose morale. They concluded that they could not wait for *Syren*, which had separated from *Intrepid* after the storm and agreed to meet later that evening with boats to help screen *Intrepid*'s retreat after the mission. The winds were favorable for both a smooth entry into and exit from the harbor, and the storm clouds appeared to be a day off. Decatur ordered his crew to clear the decks and make ready for battle. That very morning Bainbridge had encoded and sent another letter to Preble about a raid with a sense of urgency to it, saying, "[H]aving had the liberty of walking out and have seen the situation she is anchored in, do conceive it would be easily effected."[25]

The men concealed themselves, mostly below, and as the sun set *Intrepid* made its approach into the harbor. Salvatore Catalano, the Italian pilot sent by Preble, had sailed the Mediterranean for decades and knew the harbor well. He guided the ketch past the reefs and rocks and through the anchored ships, the relatively massive hull of the *Philadelphia* easily visible in the reflected lights of the city. As *Intrepid* approached the frigate, a sentry called out; Catalano called back in a local language, a Mediterranean sailor's patois of mixed dialects, and talked the vessel alongside claiming to have lost his anchor in the storm the day before. *Intrepid* made fast to *Philadelphia*, and the Americans struck.[26]

The boarding party, made up of sixty of the crew, poured from below decks and scaled the side of the frigate. Midn. Richard Morris, who would later command USS *Adams* in the War of 1812 and be promoted to commodore, was the first to reach *Philadelphia*'s deck "in a spirit of gallant emulation," followed closely by Midn. Thomas MacDonough, who would become the hero of the battle of Lake Champlain. The Americans fell upon the Tripolitan guards with swords and knives, under strict orders from Decatur not to use firearms for fear of alerting the harbor's defenders. The attack went like clockwork, as each of the men went rapidly about his assigned task. The rapid strike killed twenty guards, and took a single prisoner, and the men then began setting up the combustibles. Several Tripolitans, however, escaped in a boat moored on the opposite side from *Intrepid*, or had jumped overboard; the alarm went out across the harbor, and the fortress opened fire.[27]

As the raiding party went about their business, Catalano glanced around the harbor. The winds continued to favor their escape; the tides, current, and the layout of the ships anchored in the harbor were better than he had anticipated. He found Decatur and explained that they might be able to bring the frigate safely out of the harbor after all, even without its foremast and with only a skeleton crew. The lieutenant, however, had his orders. *Philadelphia*—the frigate his father had commissioned—had to be destroyed. Decatur ordered the fires set in the storerooms, gun room, cockpit, and berth deck. The lieutenant ordered the men back on board *Intrepid* as cannon shot from the fortress flew overhead. The rapidly spreading flames poured from the hatches and ports as Decatur himself crossed back to the ketch. When he cast off, the fire had begun to climb the frigate's rigging. Under Catalano's orders to the helm, *Intrepid* began to make its way to the channel, firing its four guns and muskets back at the harbor fort.[28]

As *Intrepid* cleared the harbor it was joined by *Syren*'s boats, which had followed the original plan despite *Intrepid*'s earlier-than-expected attack. They helped tow the ketch past the rocks and covered the escape. *Philadelphia*, engulfed in flames and its cable burned through, drifted through the harbor and finally came to rest against the fortress. In the confusion of the flames and cannon fire in the harbor there was little attempt to chase the escaping Americans. *Syren*'s men augmented the crew of the *Intrepid* and transferred fresh stores as the second gale began to close in. Together the American ships weathered the storm and began the long transit back to Syracuse.[29]

The excitement over the successful raid into Tripoli harbor was palpable. When word reached the United States, Congress immediately promoted Decatur to captain at the request of Secretary Smith. Back in the Mediterranean Midn. Ralph Izzard, who participated in the attack, wrote to his mother that "we are astonishing the folks in these parts . . . the Commodore has new schemes in his head. . . . I expect to go to Naples shortly and then we will have hot work off Tripoli." The young officer, who was soon promoted to lieutenant, was right. The mission into Tripoli served as an operational turning point for the American squadron. Preble's efforts to obtain gunboats for attacks on the harbor and the pasha's ships began to bear fruit. While American consul James Cathcart in Leghorn worked on the possibility of building gunboats for the squadron, Preble applied directly to the Kingdom of the Two Sicilies to lease or borrow some of their small combatants, since there were "a great number of very fine boats at Palermo and Messina." The news of Decatur's success arrived in the Italian court at the same time they were considering the request, likely contributing to their decision to help the Americans. By the end of the spring, when Preble visited Messina and Naples on board *Constitution*, King Ferdinand II had agreed to deliver six gunboats, two mortar boats, and munitions and crews as a loan to the American squadron.[30]

Having maintained a blockade for years, patrolled the Mediterranean, and made captures of blockade runners and corsairs, the character of the American operations against Tripoli changed. The squadron began offensive operations in earnest following the successful raid to destroy *Philadelphia*. On 25 July, Commodore Preble approached Tripoli harbor with an assault force made up of *Constitution*, the brigs *Argus*, *Syren*, and *Scourge*, the schooners *Enterprise*, *Vixen*, and *Nautilus*, two bomb ketches, and six gunboats with over a thousand officers and men. Preble's intelligence told him that Tripoli was protected by its fortress walls, batteries made of 115 guns, with nineteen gunboats, two galleys, two schooners, and a brig. The commodore envisioned the attack on the harbor not as a struggle to take the city, but instead and effort to destroy the pasha's means of defense by eliminating his naval forces. Once the city was defenseless, the Americans could tighten the blockade or begin regular and easy bombardment of the city, thus offering Preble and his ships the chance at greater strategic effects. The shallows of the harbor and the channel made it impossible for *Constitution* to get in close, which left the main fighting to the shallow draft vessels, with the frigate's heavy battery waiting to provide fire support if Tripolitan gunboats came in range.[31]

As July came to a close, and Preble waited for the right conditions to attack, a gale blew off the North African coast and sent the American squadron back into the blue water to weather the storm. The squadron returned to its close position on 3 August and at noon Preble signaled his commanders to come within hailing distance. A group of Tripolitan gunboats had ventured out of the channel and into the open water beyond the rocks that protected the harbor. Preble intended to make them pay for their boldness. After days of planning and multiple conferences with his junior officers, he ordered them to beat to quarters and prepare to attack the formation of gunboats and the harbor. At 1430 *Constitution* hoisted the signal to commence the assault and the small ships sailed in for Tripoli. The bomb ketches commenced action fifteen minutes later by hurling shells into the city itself. The two divisions of gunboats, under the command of Stephen Decatur and Master Commandant Richard Somers advanced on the Tripolitan gunboats and a heavy and general battle ensued.[32]

At 1600, as the sun sank toward the horizon and winds began to shift, Preble signaled the retreat from *Constitution*'s yardarm. He wrote in his report to the secretary that "our grape shot made havoc among their men, not only on board their shipping, but on shore" and concluded, "the enemy must have suffered very much in killed and wounded." The American attack sank three Tripolitan gunboats outright, raked several more with grapeshot and musket fire which cleared their decks of any living resistance, and took three as prizes. The Americans did not lose a single ship or boat, had thirteen wounded, and a single sailor killed in action. The death was Lt. James Decatur, the younger brother of the now-celebrated Stephen and a well-regarded officer in his own right, which moved the entire American force toward a desire for vengeance.[33]

On 7 August, Preble received a letter from the French consul in Tripoli suggesting that the pasha was ready to negotiate. However, when the Americans sent a boat into the harbor under a white flag, the palace did not reciprocate with their own flag of truce and the boat withdrew. Preble, incensed, elected to mount another assault. With his gunboats reinforced by the three prizes taken the week before and manned with sailors borrowed from the ships of the squadron, the two divisions under Decatur and Somers launched a second assault under sail and sweeps. Likely because of the gunboat losses the Tripolitans had suffered in the first attack, the enemy boats remained deep in their harbor and Decatur and Somers took their boats up point-blank to Tripoli's western shore batteries. While the first attack had had few consequences for the shore defenses, the second assault appeared to have a greater effect on the batteries by dismounting guns and killing gun crews. However, this increased result ashore also came with a greater loss. Firing hot shot, one of the Tripolitan gun crews hit *Gunboat 9*, one of the prize boats under the command of Lt. James Caldwell. The superheated ball ripped through the boat's magazine in the stern and it exploded, taking with it ten Americans including the commander. The handful of crew remaining continued to return fire from their bow gun under the command of Midn. Robert Spence, as the hull descended into the water. The other boats of the flotilla rescued Spence and ten other survivors. The attackers retreated on the commodore's signal a little after 1700 as the wind began to shift directions.[34]

During the second attack on Tripoli, the corvette *John Adams* appeared on the horizon and joined the American squadron. Fresh from the United States, Master Commandant Isaac Chauncey brought dispatches from Washington, which included the formal commission for Stephen Decatur's promotion to captain. However, the dispatches also brought bad news for Preble. The Navy Department was dispatching a new squadron, under Capt. Samuel Barron, with reinforcements of more heavy frigates, not long after *John Adams* sailed. Barron was senior to Preble and would take command in Mediterranean.[35]

On 9 August, the squadron lay offshore resupplying the gunboats and bomb ketches with ammunition and stores in order to prepare for another attack. On the following day, Preble sent a boat into the harbor under a flag of truce with letters for Bainbridge and the American prisoners of war. The boat returned with a letter from the pasha offering to negotiate a peace. Tripoli would accept a ransom of five hundred dollars a head for the prisoners, end the war, and give up the demand for increased tribute. It was a significant concession, but Preble thought it was a sign that he was approaching complete victory. In total, the pasha's offer would have cost nearly $200,000. Preble made a counteroffer of $100,000 total ransom for the prisoners, a $10,000 consular gift, and no tribute. Karamanli rejected the counteroffer and the squadron continued preparing for a final attack that Preble planned to execute when Barron arrived with the reinforcements.[36]

Despite Isaac Chauncey's report that the new squadron was leaving within days of his own departure from the American coast, days stretched into weeks. In his diary, Preble recorded day after day his readiness to attack and concerns that a shift in the weather that normally came with the end of summer would push his assault force further offshore and make a new operation impossible. On 14 August, the commodore decided he could not wait for the arrival of reinforcements and prepared for a night attack. But the winds came up and raised a heavy sea that made the small-boat operations "imprudent." Unfavorable winds three days later halted a second night attempt. Finally, on 24 August, the conditions were right and the American ships launched a night gunboat assault, but again failed to force the pasha to surrender. On the evening of 27 August, *Constitution* and the squadron bombarded the city. While they appeared to wreak havoc on the shore defenses, it still had no effect on the pasha's willingness to concede.[37]

In July, Preble had ordered *Intrepid* converted to a hospital ship to support the attacks on Tripoli over the summer, and in between the combat operations the vessel carried supplies to help keep the ships on station. In the middle of August, *Intrepid* had sailed from Syracuse with a cargo of stores to help resupply the squadron off Tripoli. The crew filled the hold and deck with water casks, eggs, sheep, pigs, and fresh fruit and vegetables. After the ketch arrived with the squadron, Preble placed her under the command of Midn. Joseph Israel as the ship moved around the squadron distributing supplies. When the ships and gunboats moved forward for the bombardment on 27 August, *Intrepid* remained offshore with *John Adams*, waiting to assist casualties. But there was more in store for the captured ketch. Following the failure of the bombardments at the end of August to force a resolution, Preble ordered *Intrepid* alongside *Syren* and began outfitting the small ship for a new irregular mission as an "infernal": an explosive-laden fireship, which a crew would sail into the harbor to destroy the remainder of the pasha's naval defenses.[38]

By the start of the nineteenth century, fireships had developed a long and feared history in naval warfare as an irregular weapon. Although they were more commonly used to attack larger enemy warships in the littorals, Preble and his officers devised a way to turn *Intrepid* into a giant unmanned bomb, which they could send deep into the harbor. The target was the remaining Tripolitan gunboats, which lay moored together at the base of the city's walls. Preble went on board the ketch to inspect her and look over the planned changes to the ship. The officers brought carpenters from across the squadron together to make the modifications needed to pack the hold with a hundred barrels of gunpowder and over a hundred explosive shells. They worked for three days preparing her for the mission. The carpenters boarded over the forward section of the hold and filled the compartment with loose powder, then covered the top with the explosive shells. They placed more powder in the aft hold in casks and added ballast iron, which the massive explosion would turn into shrapnel. A fuse system was devised with wood shavings and turpentine around it that would make any boarding party assume the ship was about to blow, even if there was time left on the fuse.[39]

Despite Midshipman Israel's nominal command of the ship, Preble hand-selected the crew that would sail her into the harbor. Richard Somers had recently been promoted from lieutenant to master commandant. Throughout the summer, he had been leading the second division of gunboats alongside Decatur's command of the first division. The squadron's wardrooms lionized Somers nearly as much as Decatur. He had commanded *Nautilus* prior to the summer's gunboat campaign and had a reputation as a good skipper. The two men were close friends, having served together under Capt. John Barry on board *United States* during the Quasi-War. Somers was actually senior to Decatur on the lieutenant rolls by one day, until the post-*Philadelphia* promotion to captain arrived. He had clearly demonstrated his leadership and combat skill, keeping up with the hero of the raid on *Philadelphia* in attack after attack. The commodore selected him to lead *Intrepid*'s final mission.

Preble added Henry Wadsworth, a newly promoted lieutenant, as the second in command. To round out the crew for *Intrepid*'s raid, the commodore assigned sailors from Somers's command, *Nautilus*, and from *Constitution*, where Wadsworth served. At the last moment, Midshipman Israel joined the crew with orders from the commodore. Israel stayed on board, whether under the orders of Preble or by talking Somers into letting him participate in the mission with "his" ship remains unclear.[40]

The plan was relatively simple. The carpenters ran fuses into the holds that were set to burn for fifteen minutes. Somers and his small crew of twelve men would sail *Intrepid* into the harbor in the dark of night. Once they were past the rocks and in the channel, and with a direct line to run the ship up into the enemy's gunboats against the main fortress's harbor wall, they would tie off the rudder and the rigging so the ship would sail herself. The officers had selected the two fastest rowing boats in the squadron, including the best cutter from the newly arrived *John Adams*, to take with them. After the crew lit the fuses, they would abandon ship, as *Intrepid* continued westward into the harbor. Rowing hard in the picked boats, they would make their escape back to the squadron. *Intrepid* would continue her final minutes alone, crashing into the fortress and exploding.[41]

On 2 and 3 September, *Intrepid* got under way and approached the harbor, but the winds were not right for the attempt. On the third night, after a bombardment by some of the squadron's ships, *Intrepid* sailed again, escorted in shore by the schooners *Argus*, *Vixen*, and *Nautilus*. The winds were favorable, and Somers and the crew headed into the channel. *Nautilus* followed, taking the role that *Syren* had played in Decatur's raid. Lt. George Washington Reed and Lt. Charles Ridgely, Somers's lieutenants who were commanding the schooner in his place, took her in as close as they could before they feared being spotted from the harbor, then bore off to wait for the escaping rowboats.[42]

Intrepid sailed westward along the channel and into Tripoli's harbor. As Ridgely watched through his night glass on the deck of *Nautilus*, the white canvas of his skipper's sail, full with an easy breeze out of the east, ghosted into the darkened port. It appears, however, that the Tripolitans had spotted *Intrepid*. Three shots rang out from a battery facing the harbor's entrance. There was a brief moment of silence, and the Americans watching the operation breathed a sigh in the quiet. The timepieces on board the squadron approached ten o'clock. Suddenly, well before the expected time for the detonation, the air and water around Tripoli filled with the light and crashing madness of a massive explosion. Ridgely could see the masts of the infernal ketch rocketed into the sky as the harbor was lit up bright as day. Preble wrote, "[T]he shrieks of the inhabitants informed us that the town was thrown into the greatest terror." Shells and shrapnel flung from the explosion began raining down in all directions. But *Intrepid* had not made it to her intended point of attack, exploding in the outer section of the harbor. While the attack appeared to strike fear into the city and silenced the guns of the enemy for the rest of the night, the material damage was negligible.[43]

Nautilus and the other schooners patrolled the entrance to the channel, desperately hoping to spot *John Adams*'s green cutter and the raiders' other boat. Preble wrote, "the whole squadron waited, with the utmost anxiety, to learn the fate of the adventurers . . . but waited in vain." The rescue schooners remained under the guns of Fort English, Tripoli's easternmost gun emplacement, until well after the sun had risen and revealed their position. Finally, around eight in the morning, the ships hauled off and set sail to meet up with the rest of the squadron.[44]

What actually happened to the thirteen men on board *Intrepid* remains a mystery. Commodore Preble, Lieutenant Ridgely, and probably every other officer and sailor in the squadron had their own theory. Because one of the Tripolitan gunboats appeared to be missing the following morning, some believed it had tried to attack and board *Intrepid* after the cannon fire from the shore battery. Others thought Somers and his men, fearing not only their own capture after being spotted but also the capture of so much valuable gunpowder, sacrificed themselves. Preble told Secretary Smith that he believed four enemy gunboats had attacked *Intrepid*, Midshipman Spence wrote to his mother that there were two. Both explained that their story was mostly conjecture. Even the American prisoners in the city added to the speculation, later relating that there had been no reports of Tripolitan injuries during the attack, which suggested there were no gunboats alongside *Intrepid* when it blew up. In fact, there had been so little damage that *Philadelphia*'s surgeon said the pasha's court held a prayer of thanksgiving. The Tripolitans recovered ten bodies on the rocks and floating in the harbor the following morning.[45]

As the schooners joined the squadron offshore on the morning of 5 September, the seas began to rise. Preble's crews continued re-arming the gunboats for another attack, but the winds shifted out of the north-northeast and began pushing a heavy swell toward shore. The weather that Preble had been worrying about for the past few weeks was setting in. He ordered the men to reverse their work, taking the heavy shot and shell out of the gunboats and bomb ketches and stowing it back on board the larger *Constitution* and *John Adams*, which would ride the heavy seas more safely. Over the next two days, as the crews worked to make their ships safe, the weather continued to deteriorate. Preble took stock of the ammunition and stores that he had left. There was barely enough to maintain three ships for the blockade, never mind the entire force he had on the Tripolitan coast. Barron still had not arrived with reinforcements or resupply and Preble decided that his window to force the pasha's hand had closed. On 7 September, he ordered *John Adams*, *Syren*, *Nautilus*, and *Enterprise* to take the gunboats and bomb ketches in tow and head for the base at Syracuse. He took the remaining ammunition and stores from the departing ships and continued on station with *Constitution*, *Argus*, and *Vixen* to hold the blockade.[46]

On the afternoon of 10 September, *Argus* was chasing a blockade runner who had attempted to escape the harbor when lookouts spotted two sails. It was the frigates *President* and *Constellation*, finally arriving to join the blockaders. *Argus* shortened sail to keep company and pass messages with the frigates, and sent signals to *Constitution* announcing the arrival of the reinforcements. The blockaders joined up with the new frigates and the captains of the ships joined Barron on board *President* for a conference on 10 September. Barron had been "mortified extremely by the contrary winds which has [*sic*] lengthen'd our passage in an uncommon degree" and stopped at Malta for water and stores before heading to meet Preble off Tripoli. At the conference of the ships' captains Preble began briefing his replacement on what the squadron had accomplished over the summer, and they continued in private meetings. Secretary Smith's orders expected Preble to remain in the squadron, giving up his position as commodore but staying on as captain of *Constitution*. Preble, however, had other ideas. He was tired and after such successful combat operations over the summer he was not about to take a position as third in command of the squadron. Capt. John Rodgers had sailed with Barron, and he was also senior to Preble. Answering to Barron was one thing, but Preble and Rodgers had a long-standing animosity and Preble had sworn never to serve under his command.[47]

Barron agreed with Preble when he told the new commodore that he intended to go home. He also listened to Preble's advice and let the departing commodore order the newly promoted Capt. Stephen Decatur to command of *Constitution* as one of his final acts in command of the squadron. After hearing about the success of the summer gunboat attacks and harbor bombardment, Barron asked Preble for one thing before he sailed for home: return to King Ferdinand of the Kingdom of the Two Sicilies and attempt to get more gunboats for the next year's operations. The squadron had returned the boats used in 1804 to their owner when they made it back to Syracuse, and the Americans clearly needed more small ships. Preble agreed to work his naval diplomacy one last time, but was unsuccessful. Ferdinand was pressured from somewhere else to withhold support from the Americans. Preble suspected the French. Before sailing for America, he sent one last letter to Barron suggesting buying more gunboats in Malta, where the relationship with Governor Sir Alexander Ball had been strong.[48]

The American squadron never obtained more gunboats from allies in the Mediterranean. William Eaton, the former consul in Tunis, had returned to the Maghreb with Barron's reinforcements and orders to explore the possibility of an overland campaign in support of the pasha's older brother and his rightful claim to the throne. The success of that campaign, and the taking of the city of Derna in April 1805, put enough pressure on Yusuf Karamanli that he acceded to terms in June. Eaton was beside himself, forced to abandon the brother Hamet Karamanli, but the war was over.[49]

The Barbary War has taken on a strange mythic place in American naval history. Despite the fact that it was the second war that the U.S. Navy fought after its refounding in 1798, many historians and authors continue to label it the "birth" of the U.S. Navy and Marine Corps. This includes classically trained and academic historians as well as the more recent crop of journalists and political commentators that have rediscovered what they commonly call the "forgotten" war. While the conflict with Tripoli clearly was not the start, or birth, of the American sea services, it did demonstrate the sustained importance of naval irregular warfare and maritime raiding in the development of American naval operations. In doing so, it continued the trajectory that carried American sea power through the founding of the service and the Quasi-War, a trajectory that remained well in line with the operational experience of John Paul Jones and with the Revolution.[50]

Edward Preble had a clear sense of the balance that successful operations in the Mediterranean required. Unlike his predecessors, who focused on blue-water missions of convoy escort, patrol, and blockade, Preble understood that bringing the war to a positive result required something more. Like Silas Talbot in the West Indies, Preble's own experience on board *Winthrop* during the Revolution made him predisposed to consider and value irregular operations and shallow-water combat as important elements of his overall campaign plan. He also demonstrated the same kind of trust in his subordinate commanders and junior officers as Talbot had, which had led to irregular successes in the past. Where Talbot had Hull and Porter, Preble turned to Decatur and Somers. While it is perhaps an apocryphal reference, maybe the invention of a prior historian or based in the oral tradition, the claim that Nelson himself remarked that the attack on *Philadelphia* was the "most daring act of the age" has the ring of truth, despite its dubious provenance. Preble actively managed the daring irregular acts of his lieutenants with the continued balancing of the blue-water blockade and patrols.[51]

During the Revolution, Paul Jones's personal knowledge of Whitehaven and St. Mary's, and the information provided by the cooperative Irish fishermen, made his raiding operations possible. Two decades later, the cooperation of Thomas Sandford assisted in Hull and Carmick's mission at Puerto Plata. In the Mediterranean, Preble followed suit by working to maintain critical partnerships. His employment of the pasha's former doctor Pietro Corcillo as a surgeon's mate on board *Constitution* helped confirm his suspicions of *Mastico* when Decatur's crew first boarded the ketch. In the raid into the harbor on board the rechristened *Intrepid*, the local knowledge and help of Salvatore Catalano proved invaluable. Added to this tactical-level partnership were Preble's diplomatic efforts in the alliance with King Ferdinand of Sicily, his continued work with the British and positive relationship with the Royal Navy, and even his ability to work with the foreign consuls in Tripoli—all indicate an officer with diplomatic skill. Partnerships remained a continued part of success in the naval irregular warfare missions in the Barbary conflict.

Finally, Preble's relationship with King Ferdinand again raises the importance of fleet constitution and ship types. Preble, like Talbot and Truxton and the other West Indies commodores, found himself in desperate need of more small combatants and gunboats. The heavy frigates that sailed with the American squadron brought a significant amount of firepower for bombardment but they also had very real liabilities. This played out dramatically as Bainbridge chased his quarry toward Tripoli harbor on board *Philadelphia* in 1803, heading further inshore than he should have. *Philadelphia*'s grounding and capture likely would not have happened had *Vixen* remained close to do the inshore work. However, the squadron did not have enough small ships and Bainbridge had sent *Vixen* to patrol to the east. After his capture, the disgraced captain immediately recognized that it had been a mistake.[52] Preble clearly made an effort to maintain the pairing of his heavy frigates with small combatants, himself cruising with Decatur and *Enterprise* in late 1803 and then again keeping *Argus* and *Enterprise* with him after the summer of 1804.

Like the commodores of the West Indies, he took captured ships into American service to build the size of his small combatant force, including *Intrepid* and *Scourge*. Added to this was his turn to King Ferdinand for the gunboats and bomb ketches he needed to conduct the attacks on Tripoli. For the second conflict in a decade, the U.S. Navy's focus on heavy frigates and blue-water vessels left the operational commander with an unbalanced fleet. Small ships in the Mediterranean, like the West Indies, proved themselves indispensable for both conventional missions like blockade and convoy, but also the irregular operations and raiding missions that proved to be turning points in the conflict.

Following the conclusion of the Barbary War, the debates over the size of and need for a United States Navy returned in Washington, D.C. Like the weeks following the end of the Quasi-War, a strong alliance of antinavalists began making the case that a naval force was more trouble than it was worth. This time, however, they had a sympathetic president and executive branch. Thomas Jefferson, despite his long support for aggressive naval action in the Mediterranean and execution of the war, agreed with his Republican colleagues and began dismantling the navy. In the place of frigates, sloops, and schooners, the Jeffersonian naval policy focused instead on gunboats for coastal defense. Ironically, the pendulum swung out of balance in the opposite direction, building a large number of small combatants while ignoring the need for a balanced naval force that included blue-water ships. It was another seven years before the risks of that policy became apparent as the nation went to war with the world's maritime superpower, Great Britain.

CHAPTER 4

RAIDING ON THE LAKES
1812–1814

Following the end of the First Barbary War, the United States embarked on a period of naval retrenchment and reduction. The political interests that historian Craig Symonds characterized as the antinavalists came to power in Washington, and they replaced efforts to build an oceanic and deployable blue-water force with a focus on coastal defense. This era became known as the Jeffersonian gunboat era. The Navy Department dismissed crews and released officers from the service, and lack of funds caused them to place many of the warships in ordinary or to sell them outright. Friction with the powers of Europe continued, particularly with the British as their war with Napoleon intensified. Yet, while impressment of American sailors on the high seas continued, and incidents like the HMS *Leopard*'s 1807 attack on the American frigate *Chesapeake* raised outcries in the American press, Congress continued to resist the enlargement of the navy.[1] While some historians commonly relate the drive to war in the spring of 1812 through the lens of "free trade and sailor's rights" and attribute it to maritime interests, politically it was more than sailors' outrage that opened the door to war. Western politicians looking for territorial expansion into British-held Canada, conflict over trade restrictions enforced by the British during their war with France, British support for Native Americans in the west, and numerous minor issues contributed as well.[2]

When Congress declared war in June of 1812, the U.S. Navy was not prepared. In the months preceding the declaration, President James Madison's cabinet had even debated not using the navy in the war at all. It was not until 21 May, less than a month before Congress opened hostilities with Great Britain, that Secretary of the Navy Paul Hamilton first asked his senior captains how the navy might prepare for war, and what strategy they might pursue.[3] Capt. John Rodgers replied with a strategic discussion that was focused almost entirely on the Atlantic and blue water.[4] The traditional historiography of naval operations in the War of 1812 starts from the same assumptions as Rodgers's view of naval strategy for the war, focused on blue water. From Rodgers's own Atlantic squadron cruises to the squadron engagements on the Great Lakes, from the high seas frigate duels to the Royal Navy's choking blockade of the later part of the war, the traditional narrative of ship-versus-ship combat and conventional naval operations has dominated the scholarship on the conflict. Yet between June 1812 and January 1815, naval officers and sailors were just as likely to participate in naval irregular warfare as in conventional blue-water battles. From coastal raids and cutting-out expeditions to partisan attacks and the use of new technology, Americans conducted both conventional and irregular operations throughout the conflict.[5]

Examples of these kinds of irregular operations span the entire war, from the early fall of 1812 through the early days of 1815 before word of the Treaty of Ghent reached the United States. They also took place over a wide variety of geographies. Sailing Master John Percival, who would earn his commission late in the war and eventually command *Constitution* during an around-the-world deployment in the 1830s, was lauded for a small-boat attack on the British off New York that resulted in the capture of the sloop *Eagle*.[6] Lt. Lawrence Kearney, who would command counterpiracy missions in the 1820s and go on to play a central role in American naval diplomacy in China, led a cutting-out of British boats near Savannah directly under the guns of the frigate HMS *Hebrus*.[7] The next chapter will recount efforts to sink British warships through covert attacks in nearly every theater of the war, including Long Island Sound, the Chesapeake Bay, and the Great Lakes. This chapter focuses on the reconnaissance, raiding, and cutting-out operations conducted in the northern campaign of the war on Lake Ontario and the St. Lawrence River, and on the Niagara River, which connects Ontario to Lake Erie.

The conventional story of the lakes campaigns is of the victorious Americans in the squadron engagements at Put-in-Bay on Lake Erie and Plattsburgh on Lake Champlain, and the naval arms race on Lake Ontario.[8] But the very first success of American naval forces in the northern theater was not made up of dueling broadsides or chasing squadrons, but instead was a cutting-out expedition led by a lieutenant on the Niagara River. From that operation, through raiding on transports and supply lines to attacks on shipyards and ships under construction, the story of the campaigns on the lakes is incomplete without an examination of how small boats and daring sailors impacted the struggle for maritime supremacy and the balance of power. The naval history of the lakes in the War of 1812 provides valuable historical examples to consider when expanding the aperture of our study of naval operations to include the history of naval irregular warfare.

Early American strategy in the War of 1812, which involved operations against the Canadian frontier in order to capture enough territory to force the British into a negotiation, overlooked a significant problem: the Americans had almost no naval presence on the Great Lakes. As all three governments, the British and American and the provincial Canadian, prepared for war, it took a great deal of time to put themselves on an operational footing. The defeat of the American force under Brig. Gen. William Hull at Detroit two months into the war dashed American hopes for a quick and relatively easy conflict. Twice, Hull had asked the president and his cabinet to establish a naval force on the lakes to support his campaign. He made the case that a naval force should be formed starting with the converted brig *Adams* already on Lake Erie, but also with a new sloop of war carrying twenty guns and smaller vessels. But the cabinet denied his request because it was inexpedient and because of the costs involved.[9]

After Hull's defeat, the autumn brought the combatants closer to the end of the fighting season ashore and the cruising season afloat, and the Americans realized that naval establishments on the lakes would be required. The newly appointed commodore in command of the lakes, Isaac Chauncey, assigned Lt. Jesse Elliott to begin building facilities and capability at Black Rock, on the Erie end of the Niagara River, while he headed for Lake Ontario. The small Canadian garrison known as Fort Erie lay across the Niagara River from the navy's new base at Black Rock. As soon as they found themselves at war, the British accelerated the completion of some of the defensive positions which had been under intermittent construction since the end of the French and Indian War, and spread their troops thinly across the frontier. Across from the Americans at Black Rock, they built three small batteries but left the rest of the defensive works unfinished because of "want of means."[10]

Jesse Elliott was thirty years old and had been appointed an acting lieutenant in 1810. He trailed behind many of his peers on the navy's promotion lists. Born in Maryland in 1782, when he was nine years old his father died during fighting in Maj. Gen. Anthony Wayne's campaign against the Native Americans in Ohio. His mother wanted young Jesse to study law and managed to keep him in school until he was eighteen, but Elliott desired adventure and applied for a midshipman's appointment. He first served on board *Essex* and saw his first action late in the First Barbary War. When the War of 1812 broke out, Acting Lieutenant Elliott was looking to make a name for himself and his assignment to Lake Erie offered the opportunity he needed.[11] Elliott began establishing his facilities and started construction of two schooners and six gunboats to form a naval force on Lake Erie. He also began fitting out and arming several small merchant ships, which the Navy Department had purchased to accelerate the building of the force.

Elliott found—because of the lack of supplies, few skilled shipwrights in the area, and a lack of sailors to man ships—that his squadron was going to be a slow to take shape.[12]

On 6 October, dispatches arrived for Elliott that reported two British vessels sailing for the anchorage under the British guns at Fort Erie. The morning of 8 October found *Detroit*, the 14-gun brig formerly named *Adams* and captured by the British during Hull's surrender at Detroit, anchored in front of the British position. At her side was the 85-ton brig *Caledonia*, rated at three guns but reportedly carrying only two. The ships carried arms and prisoners captured at Detroit, as well as a cargo of furs from *Caledonia*'s owners. Elliott watched the ships offloading their cargo to Fort Erie and received the first positive report that he had seen in weeks: the officers and men whom Commodore Chauncey had promised him were on the wilderness roads not far from Black Rock. He dispatched a rider with instructions for the column to hurry, because he had "determined to make an attack" on the anchored British ships.[13]

The men who marched from Lake Ontario appeared at Black Rock around noon and Elliott ordered them to rest for the attack he was planning for that evening. The sailors arrived almost entirely unarmed, with only twenty pistols and no cutlasses or axes. Elliott turned to the U.S. Army commander at Black Rock, Brig. Gen. Alexander Smyth, and the commander of the frontier militia Maj. Gen. Amos Hall. The army commanders listened to Elliott's plan and offered him the small arms his sailors needed from their magazines, and General Smyth placed fifty regulars under Elliott's command for the expedition.[14] As the Americans prepared, Capt. Harris Hickman, a U.S. Army officer captured at Detroit, crossed the river from Fort Erie after the British commanders granted him parole. He reported everything he had seen in the British positions and on the ships, including the manning of the vessels, what cargo they carried, and the fact that they were essentially in a "defenceless state."[15] Elliott had two long boats, capable of carrying fifty men each, hidden in Buffalo Creek with quick access to the Niagara River.

After midnight on the morning of 9 October, Elliott and the joint force of sailors and army regulars boarded their boats and set off down Buffalo Creek to cross the mouth of the Niagara. With the current against them, it took the men two hours of rowing to approach their targets. At 0300, they arrived alongside and caught the British entirely unprepared. Pouring onto the quiet decks of *Detroit* and *Caledonia*, they subdued their skeleton crews. Within ten minutes of the first man over the sides, the Americans had secured their prisoners and the sailors sheeted the topsails and slipped the anchors. There was very little wind in the early morning hours, and the current, which the Americans had battled with their oars to get in position to attack, was now pushing them away from their destination in Black Rock harbor. This pushed the ships directly into the guns at Fort Erie. The British commanders received a report that three hundred to four hundred men had attacked the two ships and ordered an alarm. The cannon mounted in the incomplete defensive positions at Fort Erie fired round, canister, and grapeshot into the two ships as they were carried past on the current. Elliott and his men worked to maintain control of *Detroit* as the crew on board *Caledonia* beached her under some American guns on the river's eastern shore. As *Detroit* continued to drift, the British began shifting their flying artillery down river to keep the ship under fire. Elliott, having received a mistaken report that another British ship was sailing to the river's mouth from the lake, dropped his anchor and ordered all of the guns hauled over to bear on the Canadian shore.[16]

Caledonia was relatively safe under the American guns, but the sailors on board *Detroit* engaged in a continuous battle with the British fixed batteries and artillery positions that had moved down river. Elliott attempted to send a heavy line ashore to pull the brig over to the eastern bank, but was foiled by the current. With all fourteen guns hauled over and firing on the British positions the Americans exhausted their powder and shot. In an effort to get out from under the heavy cannon at the fort, they cut their anchor cable and again began to drift. Heading stern first down the river, they ran aground on Squaw Island. Elliott loaded his prisoners and men into the boat they had used for the assault and a skiff they found on board and abandoned ship. He reported that "she had received twelve shot of large size in her bends[,] her sails in ribbons, and her rigging all cut to pieces."[17]

With her cannon still on board, and two hundred muskets in her hold, the grounded *Detroit* remained valuable to both the Americans and the British. British forces twice sent boats to board her, but fire from the American shore drove them away. Throughout the day cannon from both sides of the river made any attempt at recovery fruitless. Elliott realized that, with the battering it had taken from both sides, he would be unable to salvage the ship. As the evening approached, he was determined to destroy it. After nightfall, a small group of Americans slipped on board, salvaged some of the guns, and set the ship on fire.[18]

Despite the loss of *Detroit*, the expedition was an important success for the U.S. Navy. The first naval battle on the lakes had ended in the Americans' favor. Elliott desperately needed the cannon recovered before the destruction of *Detroit* to arm the civilian ships he was converting. *Caledonia* carried $200,000 worth of furs, which the owners were trying to ship to England, and the cargo was condemned as part of the prize. After the prize money had been calculated and assigned, Elliott took *Caledonia* into the U.S. Navy.[19] Eventually, Lt. Daniel Turner took command of the small ship, and her two long guns proved invaluable the following summer at Put-in-Bay; she was the only ship with the range to strike back at the British as they pummeled Master Commandant Oliver Hazard Perry's flagship *Lawrence*. *Caledonia* would also sail with the missions to Lake Huron and Lake Superior in 1814 before the department sold the schooner at Erie, Pennsylvania, at the close of the war.[20] Commodore Chauncey wrote to Secretary Hamilton that "Lieutenant Elliott deserves much praise" and he "had no particular orders from me" and had taken his own initiative. Chauncey rightfully predicted to the secretary that Elliott's success marked the beginning of American ascendancy on Lake Erie.[21]

The success at Black Rock set important precedents for naval irregular warfare on the Lakes. In keeping with the traditions that stretched back through the first decades of the American naval experience, Elliott showed the importance of aggressive leadership, but also the ability to balance that with a good assessment of risk as shown in his decision to abandon and eventually destroy *Detroit* rather than risk his force to keep the ship. He also used local U.S. Army and militia resources effectively as partners, obtaining much-needed arms and ammunition as well as supporting manpower from the regulars offered by the ground force commanders. The centrality of joint operations in naval irregular warfare in the Great Lakes theater will be repeated throughout this chapter. It stemmed from good relationships between navy and army officers at the local and tactical level, even if American forces did not appear to cooperate well strategically.

Elliott's first American naval success was followed the next summer by Perry's victory against the British squadron at Put-in-Bay on Lake Erie, securing American control of the upper lake. Lake Ontario, however, was a different story. Commo. Isaac Chauncey of the United States Navy, and Commodore Sir James Lucas Yeo of the Royal Navy, maneuvered their squadrons and built ships as fast as they could on Lake Ontario. Each attempted to gain a decisive material advantage. The result was a naval arms race on the American and Canadian frontiers. For the most part the commodores husbanded their resources and acted with caution in order to build their squadrons.[22] Chauncey had an opportunity for victory over Yeo in September of 1813 but refused to divide his squadron by leaving his slower schooners behind. The British escaped from the engagement, which later earned the nickname "The Burlington Races" because they raced for the protection of the shore batteries at Burlington Heights at the western end of the lake rather than face the Americans. Limited to the speed of his slowest schooner, Chauncey's pursuit of the weakened and fleeing British squadron was hampered enough to allow the British to escape to their anchorage under the cover of their artillery.[23] The squadrons did come into contact a number of times over the years of the war, but the decisive battle never took place and each commodore blamed the other for always running from a fight.[24] From his vantage point on shore, a young Brig. Gen. Winfield Scott observed, "[T]he two naval *heroes of defeat* [*sic*] held each other a little more than at arm's length—neither being willing to risk a battle without a decided superiority in guns and men."[25]

Robert Malcomson's study of the conflict on Lake Ontario highlights the complexity of the decisions that the commodores faced, and illustrates why they made those decisions, which resulted in an apparent stalemate. However, looking past the question of squadron engagements, or even ship-on-ship duels, the naval operations on the lake included more. Action ashore by the armies of both nations, and movement of the supplies needed for the constant competition in naval construction, required the support of a fleet of small transports and coastal supply ships. These vessels worked Lake Ontario's shallows and choke points, conducting joint irregular operations for the theater's armies and navies. As Howard Chapelle observed, "[B]oth sides suffered from the difficulties of getting materials and equipment from their coast," and both navies required steady supply lines in order to support their building efforts.[26] The British went so far as to build modular components in England that the Admiralty shipped to Canada for assembly.[27] The transports, and the gunboats that offered them protection, provided ample opportunity for naval irregular warfare operations.

In July of 1813, a pair of men approached Commodore Chauncey asking to launch a privateering mission into the St. Lawrence to attack British shipping. Samuel Dixon, a lake mariner who would continue to work with American naval forces on Ontario for the rest of the war, and a local militia major named Dimock proposed to fit out a pair of schooners to raid northward. *Fox*, with a single 8-pound long gun, and *Neptune*, with a 6-pounder, loaded fifty volunteers from the regulars and militia at Sackets Harbor. With another lake mariner from Sackets Harbor named William Vaughn as the skipper of the second ship, they sailed for the mouth of the St. Lawrence on 18 July.[28]

The two American ships fell in with a British convoy on the afternoon of Monday, 19 July. The British gunboat *Spitfire*, with a single carronade, was escorting fifteen transport boats. But the Americans surprised the defenders so totally that they overwhelmed the gunboat and took control of the convoy with neither side firing a single shot. Dimock and Dixon discovered they had taken a valuable prize. It was a full convoy from Montreal to Kingston with 250 barrels of pork, 300 bags of pilot bread, ammunition, and assorted stores.[29] However, the American commanders realized that as well as having a large prize, they were also overwhelmed with 67 prisoners to guard, 15 transports, and a gunboat to control. They headed for the American side of the river and found a place to collect themselves and their prizes in Cranberry Creek, near Goose Bay on the southern side of Wellesley and Grenadier Islands. With the transports and captured gunboat anchored, the volunteers began setting out defensive positions while the commanders worked on determining the best way to get their prizes back to Sackets Harbor.[30]

A local Canadian man named John Andrews noticed the group of boats anchored near Goose Bay and reported it to the British authorities. The express rider arrived the morning of 20 July and reported the loss of the convoy and its likely location to the British commanders. The British dispatched three gunboats under the command of Royal Navy lieutenant John Scott with men from the squadron, and a fourth with men from the 100th Regiment under the command of Capt. John Martin. The four gunboats rendezvoused off Long Island (known today as Wolfe Island) east of Kingston and headed downriver. By the time they reached the vicinity of the American position, darkness was setting in and they stood off for the night. While they were waiting for the morning, a detachment from the 416th Regiment of Foot and Major Richard Frend, who took command of the operation, arrived to reinforce the British party.[31]

Before 0300 on the morning of 21 July, the British force under Frend crossed the St. Lawrence and started into Cranberry Creek, hoping to arrive at the American position as dawn broke. What the British did not count on was that the Americans had spent their spare day constructing defensive works. They had pulled their boats farther up the creek until it narrowed to the point of making maneuvering almost impossible, then they had offloaded their prizes to stow the captured materiel in the forest. They cut trees into the creek to block the British force further and constructed bulwarks along the shore made of the barrels and bags of captured stores. The British force pulled up the creek, into the ambush laid by the Americans.[32]

As the gunboats stopped at the obstructions, and British sailors began attempting to clear the felled tree trunks and brush blocking their path, Dimock and Dixon sprung their trap from the north bank of the creek. Under a steady musket fire, the British attempted to land a group of troops on the south bank to march inland and cross back over to the American position. But the way was impassable and the troops had to return to the boats, still under steady fire. The detachment looked for a place to land on the north bank but could find nothing. Meanwhile, the gunboats could only bring a single gun to bear on the American positions, and the heavily wooded area and breastworks thrown up by the defenders left that single gun, apparently without canister or grape to clear things out, all but useless. In a move that appears desperate, Lt. Richard Fawcett of the 100th Regiment led fifteen of his men into the waist- and chest-deep water, holding their weapons and ammunition over their heads, and slogged to the northern bank. On dry land, the British regulars pushed a group of American skirmishers back to their heavier defenses but could make no further progress. Frend determined that the recapture of the convoy was not worth the sacrifice that would be required to overwhelm the well-situated American defenses, re-embarked his troops, and pulled back toward the St. Lawrence. Free of the creek, the British tended to their wounded and the gunboats headed back to Kingston with four soldiers killed (including General Sir George Prevost's aide-de-camp) and twelve wounded, along with an injured midshipman and four sailors from the boat crews.[33]

While elated at their victory, Dimock and Dixon realized they were still in a precarious position with more than forty-five miles of water to cover, two-thirds of it in the St. Lawrence, to get back to Sackets Harbor. The privateers reloaded their captured supplies onto the boats and pulled back into Goose Bay. What they did not know was that word of the engagement had reached Commodore Chauncey by rider. He dispatched the small schooners *Governor Tompkins* (6-gun), *Conquest* (3-gun), and *Fair American* (2-gun) from Sackets Harbor to patrol between the mouth of the St. Lawrence and Grenadier Island, covering a portion of the escape route. The Royal Navy, however, had also deployed to block the escape with the HMS *Earl of Moira*, a 70-foot brig, (with reportedly fourteen, but more likely sixteen, guns on board), taking up a position in the mouth of the St. Lawrence to intercept the escape.[34]

The Americans sailed from Goose Bay and along the American side of the St. Lawrence past Wellesley Island. As they approached Wolfe Island and the mouth of the river, they could see *Earl of Moira* patrolling before them. Dimock ordered his ships into a line, Dixon on board *Fox* took up the rear of the column, and the Americans decided to race for the safety of the American patrol just past the heavily gunned British ship. The single Royal Navy vessel, unable to deal with the number of American vessels swarming around it, fired into the Americans but could do little to slow the race toward the open water of the lake. In the rear of the convoy, *Fox* experienced the worst of the fight, firing back at the British with her over-matched single long gun. Passing within musket shot of *Earl of Moira*, *Fox* took three hits from the warship's long 9-pound guns but avoided the carronades of her broadside. A single shot that penetrated *Fox's* magazine continued clear through without triggering the powder or causing injury. By that evening, the prizes were all in Sackets Harbor and the prisoners under guard at Watertown.[35]

Following the loss of *Spitfire* and the convoy, the British realized their gunboat operations on the St. Lawrence were a vital part of protecting the lines of communication needed for the Kingston shipyard. The nine additional gunboats proposed to General Prevost by Commodore Yeo, in response to the attack, required a minimum of 285 men. The manning requirement would be even higher if they assigned extra marines. Yeo insisted that no transport could be allowed to move up the river without an escort to defend against American raiders. On 24 July, the day that Adjutant General Edward Baynes published the report of the Cranberry Creek engagement, the commanders ordered the entire Royal Newfoundland Regiment to be broken up and assigned to the gunboats as marines, along with "a proportion of officers, sergeants, and drummers." The 100th Regiment also had seventy troops detached for duty as marines, forty assigned to the 22-gun HMS *Wolfe*, and thirty to the 24-gun HMS *Melville*. Through the rest of the 1813 cruising season, the British built their capacity to protect against raiders, and the forces needed to conduct their own raids, but no further irregular operations of significance occurred for the remainder of the year.[36]

When the summer of 1814 began, Commodore Yeo had the superior force on the lake. At the start of May, he sailed his Royal Navy squadron to Oswego, New York, to attack the American garrison. Oswego was where supplies arrived overland for the U.S. Navy from the eastern seaboard. The Americans then moved the supplies by transports north along Lake Ontario to Sackets Harbor, where Commodore Chauncey based his squadron. The British launched a successful assault on Oswego in May of 1814, capturing the American fort and spiking the cannon. Unable to hold the poorly constructed fort with their limited manpower, however, the British sailed off with the captured American schooner *Growler* (7-gun), as well as powder and supplies from the garrison. Yeo then sailed his squadron to Sackets Harbor and blockaded the Americans at their base.[37]

Even after the successful attack on Oswego by the British squadron, supplies continued to flow into the American base. The line of communication that ran from New York City up the Hudson and Mohawk Rivers, by canal and Wood Creek to Lake Oneida, and then down the Oswego River, remained the best and fastest way to move the heavy material needed for naval construction and armament. Master Commandant Melancthon Woolsey, who had commanded the outnumbered and overwhelmed defense of the fort, remained in the area after the British withdrawal and continued to manage the flow of supplies toward Sackets Harbor. With Yeo and the squadron blockading the American base, Woolsey and his men had to change their transshipment method from larger schooners to convoys of small boats, to slip past the British blockaders. Working with troops from the 1st U.S. Rifle Regiment under the command of Maj. Daniel Appling, and a group of allied Oneida Indians, Woolsey coordinated the movement of supplies.[38]

Woolsey established a deception operation to help cover his activities. He ordered a significant number of new wagons and ox teams from local suppliers and circulated the rumor that Commodore Chauncey had ordered him to send all naval supplies back to Lake Oneida until the blockade was raised. As sympathizers in the area passed the rumors to the British, Woolsey dispatched Samuel Dixon with a gig and crew for a reconnaissance of the shoreline between Oswego and Sackets Harbor. When he returned and reported that he had not seen any British forces, Woolsey and his men loaded nineteen barges with thirty cannon, ten cables, and assorted light stores. With 150 riflemen from the regiment along as guards, the convoy set off after dark, headed north along the shore. They reached Big Salmon Creek by sunrise and were joined there by the Oneida party. They had lost one of their boats in the night and decided to slip into Sandy Creek to rest and hide the convoy.[39]

Earlier that morning the HMS *Prince Regent* (56-gun) spotted the boat, which had separated in the night, while patrolling on blockade duty. Brought alongside the warship, the British officers intimidated the American crew into admitting their mission and that they were part of a larger convoy. Yeo ordered Captain Henry Popham to intercept the convoy, assigning him two gunboats and three cutters with 160 sailors and marines. A few hours later, he reconsidered the size of the force and sent Commander Francis Spilsbury with two more boats as reinforcements, including one of his personal gigs. Popham began searching the shoreline and Spilsbury caught up with him after nightfall.[40]

About two miles up Sandy Creek, Woolsey and the convoy established a defensive position and he sent a rider with a dispatch to Chauncey to update the commodore of their progress. He suspected that the British might discover the lost transport.[41] Chauncey dispatched reinforcements overland from the U.S. Army units at Sackets Harbor: dragoons and a company of light artillery with a pair of 6-pound field pieces. As darkness approached, Woolsey ordered Lt. George Pearce to take Samuel Dixon and a boat and conduct a reconnaissance back down the creek and along the shoreline. At five o'clock the next morning Pearce and his crew rushed in to the American position with news that they had spotted a British raiding force. Woolsey began to position his sailors, marines, riflemen, and Oneidas on both banks of the creek to defend the convoy.[42]

Popham and Spilsbury landed at the mouth of Sandy Creek and scouted the American position on foot. Getting close enough to count eighteen masts collected at the edge of the marsh, Popham assumed that a small number of sailors would be guarding the easy-looking target, at most reinforced with some militia. Extrapolating from the cargo in the single boat the British had already captured, Popham rightly assumed the convoy would be valuable and he determined to attack immediately.[43] The three gunboats, three cutters, and Commodore Yeo's gig headed up the creek and commenced firing toward the Americans at eight o'clock. Less than a mile from the American position, the British force stopped briefly to land their marines on the north bank and a force of sailors on the south bank under the command of Spilsbury to move alongside the gunboats. At about 1000 the British force rounded the final bend in the creek. Firing into the underbrush with their 68-pound carronade, they flushed a group of Oneidas from their position who began retreating just as the gun was disabled. As his shore parties on the flanks fought on, and with the giant carronade disable, Popham ordered the lead gunboat to rotate in position, in order to bring the 24-pound long gun at the stern to bear on the enemy.[44]

As the gunboat attempted to turn in the narrow creek, Major Appling and his riflemen thought it was the start of a British retreat. With the British boats unable to maneuver in the winding waterway, the Americans did not want them to escape and Appling ordered his men to advance. The entire American force came out of their positions and charged the British. The newly arrived dragoons and artillery troops fell in behind the advancing riflemen, Oneidas, and sailors and began pushing back the British, who were suddenly on the defensive. In ten minutes of fighting, the British were overwhelmed and, realizing that they were overmatched, the raiding force surrendered. The Americans captured 27 marines and 106 sailors, wounded 28 and killed 14 British sailors and marines, and took all the boats and arms.[45]

After the battle concluded, Master Commandant Charles Ridgely arrived with orders from Commodore Chauncey to relieve Woolsey so that he could return to Oswego. Woolsey remained for the rest of the day, helping collect and attend to both the American and British wounded and participating in the burial honors for Master's Mate Charles Hoare, the Royal Navy's only officer killed in the battle. Ridgely took charge of moving the supplies, captured raiding craft, and prisoners back to Sackets Harbor.[46] The news of the defeat struck the British blockading squadron hard, with Lieutenant John Le Couteur writing in his journal: "So much for taking boats full of men up a wooded creek with hidden enemies on each side shooting into the boats as targets. . . . Sir James [Yeo] is in a horrible rage at their imprudence."[47] With the whole force of almost two hundred men lost in the fighting, Yeo was left trying to determine how to reman his squadron.[48]

In many ways, the engagement at Sandy Creek mirrored the events at Cranberry Creek the previous autumn. Partnerships were vital to both sides in both engagements, with joint forces of naval and army units (to include militia and allied Indians) working together. The Americans also had partners with locals, as William Vaughn and Samuel Dixon again brought knowledge of the geography and hydrography of the lakes to help the American uniformed sailors. Well-organized small-boat units also played important roles and, yet again, the British faced challenges with the armaments in their gunboats and their ability to provide effective fire support to their skirmishers once they were ashore.

With his own supplies dwindling, and facing a new and dangerous manpower challenge with the loss of his crews, Yeo consulted Lieutenant General Sir Gordon Drummond, who commanded the Lakes theater, lifted the blockade of Sackets Harbor in the first week of June, and returned to his base at Kingston.[49] The American success in the irregular engagement at Sandy Creek, therefore, had a significant operational impact. In effect, it broke the blockade and forced the retreat of the British squadron, while also halting Drummond's early plans for an assault on Sackets Harbor. Chauncey's squadron, however, still lacked the supplies needed to set out in pursuit. The interruption of the American lines of communication had ensured they would remain in port for nearly two more months.[50] The interruption to the communications needed to fit out his vessels for sea made it difficult for Chauncey to exploit the irregular tactical victory at Sandy Creek into a wider operational success. However, after realizing how tenuous his own logistics were, and refreshed by the success at Sandy Creek, Chauncey looked for an opportunity to attack the British supply lines in the same way his had been challenged. He turned to an aggressive junior officer named Francis Hoyt Gregory.

In the summer of 1814, Acting Lt. Francis Gregory began a series of irregular operations on Lake Ontario that would earn him fame with his countrymen and infamy with his enemies. The son of a Connecticut merchant captain, he spent a year in the merchant marine and was impressed into the Royal Navy for a year before escaping and receiving a U.S. Navy midshipman's warrant in 1809. The young officer, who would go on to become one of the navy's first admirals in the Civil War, was already well experienced in the unconventional elements of naval warfare when he began leading operations during the War of 1812.[51] His first ship was *Revenge*, commanded by Oliver Hazard Perry, then a lieutenant. After briefly serving under Perry in 1809, Gregory was assigned to the Gulf Coast where he began to make a name for himself as a pirate hunter. The vessels of New Orleans Station patrolled the Gulf of Mexico to protect American commerce from pirates and smugglers, as well as keeping growing unrest in the Spanish colonies from endangering American interests. Gregory served onboard *Vesuvius*, a 145-ton bomb ketch with eleven guns under the command of Lt. B. F. Reed. In February of 1810, *Vesuvius* fell in with two French ships claiming to be privateers but who were operating without documentation. Reed placed Midshipman Gregory in charge of the ketch's boats, and they captured both the *Duke de Montebillo* and the *Diomede*. Gregory's first mission under fire was a successful irregular warfare operation.[52] He was then placed in command of *Gunboat No. 162*, which patrolled the Balise, as the area at the mouth of the Mississippi River was known, and made captures at the mouth of the Mississippi and in the Gulf of Mexico.

At the outbreak of the war, he was in the midst of commanding *162* as a part of a multi-ship assault on the Lafitte brothers' smuggling base on Barataria Bay.[53] The declaration of war changed the American focus; the Navy Department dispatched experienced officers from stations on the seacoast and ships in the Atlantic to the Great Lakes to build up American forces in the northern theater. Gregory's orders assigned him to Lake Ontario, and Chauncey promoted him to acting lieutenant after he proved himself while leading men ashore during the capture of Fort George and the defense of Oswego.[54]

With fresh orders from Commodore Chauncey, on 15 June 1814 Gregory set out after sunset for the St. Lawrence River with three gigs under his command. The party headed for the Thousand Islands region, along the northeastern coast of the lake at the entrance of the St. Lawrence, to hunt British transports and gunboats. The lead small boat the raiders took with them was a beautifully constructed small craft from the village of Deal in Kent, England. It was one of two cutters, likely similar to the Deal pilot boats, which Commodore Yeo had brought with him from England to serve as his personal boats on his flagship. The Americans had captured it at Sandy Creek, two weeks earlier. Gregory likely selected it for his mission because of the Deal boats' respected sailing qualities.[55] His orders were to disrupt the transport of supplies to and from the British at Kingston.[56] At the helm of the other two boats were Sailing Master William Vaughn and the privateersman Samuel Dixon, the lake mariners who had both been at Sandy Creek and who commanded *Fox* and *Neptune* on the raid of the St. Lawrence transports in 1813. They had not only proven themselves in combat and in their ability to handle small craft, but they also provided specific knowledge of the river and its geography.[57]

Gregory and his raiders landed on Tar Island and pulled their boats ashore, hiding them in the scrub undergrowth. From the makeshift observation post, the Americans spent three days watching the British transport system. They reported to Chauncey that the British had developed a systematic early-warning network on the river. For almost the entire distance between Kingston and Prescott, which was the southern half of the voyage to Montreal from the lake, the British stationed gunboats approximately six miles apart from each other, just within visual range for signaling and for rapid response. The British augmented the patrolling boats with spotters on the highest elevations on the islands in the chain. Gregory assessed that by using the combination of the boats on patrol with the intelligence and signaling network on the islands, particularly in good weather, the British "could convey menace with great expedition." While observing the British from their hidden post on the island, Gregory and his raiders let two formations of boats pass: one because it was empty and the other because the British force was too large for the three small gigs to attack.[58]

On the morning of 19 June, *Black Snake*, a British gunboat with a single 18-pound gun and twenty Royal Marines and Canadian militia on board, under the command of Captain Herman Landon of the 1st Regiment of Granville Militia, was on patrol in the Thousand Islands. *Black Snake* was 44 feet long and lugger-rigged, with twenty-two sweeps to propel her when the wind and current could not.[59] As they neared Tar Island, the British noticed a boat approaching, with the men on board waving toward them. Thinking it was lost and part of a British convoy, Landon hailed the vessel and his crew pulled at their oars to close with the boat without manning their weapons. The boat held Gregory and a few of his men. On his signal, the other American boats made their appearance and rushed *Black Snake*. According to the British report, the aggressive attack "rendered resistance too hopeless a case, to be attempted."[60]

In the struggle to board the British gunboat one Royal Marine was badly injured, but the Americans suffered no casualties. Gregory began organizing his flotilla. He manned the prize with men taken from the crews of the three gigs and assigned guards for the prisoners. In short order the four boats headed upriver, back toward American-held waters.[61] However, the British quickly recognized the loss of *Black Snake*, likely because of the early-warning system, which the raiders had discovered. Captain Charles Owen, a militia officer, was sent in pursuit with two gunboats and 150 sailors and soldiers. The British searched the Thousand Islands and the second of the two gunboats, under the command of Lieutenant Alexander Campbell of the British Army's 104th Regiment, sighted the Americans and gave chase. The British boat, with a full crew pulling at the oars, closed on the Americans. The British fired several shots from their carronade, ranging the Americans.[62]

When the shot crossed overhead, Gregory realized having his three crews divided between the oars of four boats, and guarding the prisoners, ensured the pursuing British would overtake the small American flotilla. Before the British could close any farther, Gregory moved all the prisoners to the gigs, manned his boats at full strength, and scuttled *Black Snake*. The gunboat sank rapidly and slowed the British pursuit when they approached it. By the time the British began rowing for the Americans again, Gregory and his men had pulled themselves out of range. After the Americans escaped, Captain Owen was able to raise and salvage the cannon from *Black Snake*, and some of the stores, and the salvage operation sent the hulk to Kingston for repairs.[63] The Americans rowed through the night and arrived safely at Sackets Harbor the next morning. The results of the mission impressed Commodore Chauncey, particularly with the intelligence collected. He wrote to the secretary of the navy that Gregory "is not surpassed by any of his grade for zeal, intelligence and intrepidity [*sic*]."[64]

The reconnaissance, attack on *Black Snake*, and escape by Gregory and his men demonstrate a number of the common elements of naval irregular warfare and small-unit raiding. Gregory, who already had leadership experience in irregular operations out of New Orleans, showed that he could balance his aggressive nature as a junior officer with an ability to assess risk. His decision to allow the two convoys to pass Tar Island without attacking, as well as his recognition that, no matter the potential from prize money, it was better to sink *Black Snake* and escape with the prisoners than to be caught up trying to achieve everything at once, demonstrated his risk assessment. Vaughn and Dixon yet again brought their knowledge as local lake mariners to the expedition, as well as their recent experience at Cranberry Creek, offering important partners for Gregory. And finally, the selection of Yeo's Deal gig was an effort to ensure the proper size and rigging of vessels was available for the raiders. These principles contributed to a successful operation and Commodore Chauncey recognized they were likely to work again.[65]

With the British alarmed over American attacks in the St. Lawrence, Commodore Chauncey saw an opportunity for another irregular mission farther west on Lake Ontario. The area had been a productive hunting ground for the Americans in the past. In the summer of 1813, Commodore Chauncey's brother, Lt. Wolcott Chauncey, commanding the schooner *Lady of the Lake*, captured the transport *Lady Murray* in the same area with her hold full of ammunition and powder.[66] During the winter of 1813–14, the shipyard at Kingston was working at maximum capacity to construct warships for Yeo's squadron. Captain Richard O'Conner, who had been one of Commodore Yeo's lieutenants on HMS *Confiance* before the war, was the commissioner of the shipyard at Point Fredrick, Kingston.[67] He realized that large warships alone were insufficient: the construction of gunboats and schooners to defend shipping and supply lines remained critical. In November of 1813, he looked for other locations to begin building smaller ships and he settled on the area of Presqu'ile, on the northern shore of the lake west of Kingston. He contracted with some local men, who constructed one of the many small frontier shipyards that dotted Lake Ontario, to begin building gunboats.[68] By early summer of 1814, a schooner designed for fourteen guns was under construction there and was nearing completion. The area was also a busy choke point of British transport traffic and, with British defenses prioritized toward the St. Lawrence, Chauncey dispatched Gregory for another mission. Again, he took Sailing Master Vaughn and Dixon with him to help lead the raiding mission. This "plucky little party" sailed from Sackets Harbor in two boats on 1 July and crossed the lake from Oswego.[69]

Not long after their arrival off Presqu'ile, a British gunboat discovered the Americans and chased them away from an unescorted transport. Fearing that the British would send more patrols, Gregory and his men pulled their boats ashore on Nicholas Island, following the pattern they had established on the St. Lawrence. The lieutenant suspected that they had been spotted and, after dark on 4 July, he sent one of his boats toward shore. The crew captured a local civilian and returned to their island staging area. The local man told the Americans that, while their exact location and force was unknown, the local forces suspected that they were in the area. Two express riders had been dispatched to Kingston to alert the British commanders and to request reinforcement. Gregory and his men realized that their window for an attack was closing.[70]

The Americans loaded their boats and rowed for the small harbor where the British schooner sat on stocks, surrounded by houses. They slipped into the dark harbor and landed without notice. Gregory placed scouts at the edge of the nearby homes and sent the rest of his force under the schooner to set combustibles. He stationed his sentries to watch over the local population as much as he did to warn of an impending counterattack. In May of the same year an American raid on the Canadian village of Port Dover on Lake Erie had gone badly, with American troops and a band of Canadian militia who had volunteered on the American side rampaging through the community and burning most of the homes.[71] The outrage of the British and Canadian military and civilians was significant, and senior officers in the American services realized that the nature of the attack had crossed a line.[72]

Gregory's raiders inspected the ship as they set the fires and judged that it was less than two weeks from launch. They returned to their boats as the flames engulfed what they reported was a "stout, well built vessel." The local militia spotted the Americans but had pulled back into the town. Commodore Chauncey was clear in his orders to Gregory that the civilian population of Presqu'ile was not to be touched, and the combustibles were set to minimize the risk that the fire would spread from the schooner to the town. The commodore lamented in his report to Secretary of the Navy William Jones that a storehouse had been burned, though he confirmed that it contained supplies for the shipyard. He was clear to point out that when the raiders embarked their boats Gregory gave the order "without having permitted one of them to enter a house."[73]

Gregory's party returned to Sackets Harbor on the evening of 6 July. Once again, Commodore Chauncey was impressed, writing to Secretary Jones of "another brilliant achievement of Lieutenant Gregory with his brave companions." Just as Elliott changed the balance of power on Lake Erie through irregular warfare early in the war, Gregory and his raiders had affected British capabilities on Lake Ontario. The raiding party had not destroyed the schooner at Presqu'ile in traditional naval combat, and there was no hull to bring to port as a prize, so the men could not count on any prize money as reward for their efforts. In recognition of the importance of their success, however, Commodore Chauncey recommended to the secretary that the Department of the Navy provide a reward "in justice to these brave men."[74]

At the beginning of August, Commodore Chauncey sailed from Sackets Harbor on board his newest frigate, the 58-gun *Superior*, which was at that point the most powerful ship on the lake. He took his four strongest ships and blockaded the British at Kingston. Yeo had retired to finish construction of his 112-gun ship of the line, HMS *St. Lawrence*. Chauncey dispersed his smaller vessels to cruise the lake or to protect the supply lines between Oswego and Sackets Harbor. On 20 August, strong winds from the north blew the Americans off their blockading position at Kingston. When they returned, it appeared that some of the British had escaped. Chauncey incorrectly feared that his rival had slipped out and sailed up the lake to attack the American brigs, which were blockading Niagara. In need of intelligence, he again turned to Gregory. The British might have hidden themselves in the Bay of Quinte, west of Kingston. Chauncey sent Gregory with a midshipman and eight men to search the body of water on the north of the island of Fonta. On 26 August, he and his men departed from *Superior*, again using one of the Deal gigs captured from Yeo, with orders not to land anywhere and not to "run any unnecessary risque [*sic*]."[75]

The gig pulled into the Bay of Quinte and Gregory and his men began their search. They did not see the British squadron, or any warships, and put ashore briefly. From the shoreline, they spotted a raft of timber that a local man was transporting across the bay, likely to the Kingston shipyard. Gregory could not pass up an easy target, rowed into the bay, and boarded the raft. He and his men took the man prisoner and moved him to the gig, then set fire to the raft of timber. Without spotting the enemy squadron, the gig turned south and headed back toward the mouth of the bay.[76]

A pair of British barges, under the command of Lieutenant David Wing-field, lay hidden in the trees at the shoreline and watched as Gregory and his men entered the bay. After they spotted the smoke from the burning raft, and determining the hostile intent of the gig, the British boats rowed north into the bay. The Americans rounded the point that would lead them out of the bay's mouth, spotted the British barges, and turned to run. Gregory tossed his captive overboard, but Wingfield recognized it as a ploy to slow his chase. He ordered his men to throw the civilian an oar to float on, and to keep rowing. The British began a barrage of musket fire, which cut into the Americans as they tried to escape. Wingfield steadily took muskets and fired into the fleeing boat as one of his sailors reloaded them. Four of the eight American seamen were wounded and Midn. Ezekiel Hart was killed. Knowing there was no hope, Gregory surrendered.[77]

The British were quick to realize that they had captured the "Renegadoe" Gregory and to celebrate their success. Wingfield was even more surprised when he realized it was Gregory because the two men had met previously when, in 1813, Gregory captured Wingfield after the Americans took his ship *Confiance*. Commodore Yeo was personally informed and expressed his pleasure in the capture, not the least because Wingfield had retaken his personal gig along with the American raiders.[78] The barges brought the American prisoners back to *Prince Regent*, reaching the ship in Kingston harbor around midnight. The wounded were well cared for by the British and the Royal Navy buried Midshipman Hart with appropriate military honors.[79]

Gregory was moved north with other prisoners of war. In October, Commander Daniel Pring, who had been captured at the Battle of Plattsburg on Lake Champlain, sent a letter to officials in Montreal and enquired if he could be traded for Gregory, who had arrived at Quebec. The request was denied, despite Pring's relative seniority. Attempts by Commodore Chauncey to have Gregory paroled also met with failure.[80] The British considered him so dangerous that they transported him to England, where they held him until the end of the war. Upon his release, he headed south to meet up with the ships of the Mediterranean Squadron. His career continued for another fifty years, and his efforts in the West Indies against pirates in the 1820s are presented in chapter 6. He also commanded the Africa Squadron in antislaver operations, led a naval infantry force ashore during the Mexican War's Vera Cruz landings, and was one of the first officers appointed to rear admiral when Congress authorized the rank in 1862.[81]

The naval operations on Lake Ontario and the Niagara River exhibit a number of the principles seen in the examples of naval irregular warfare examined in the previous chapters. As an operational element of war, irregular warfare might be best considered as a contributing part of a greater overall strategy, and not an end in itself. While some of the examples from this chapter received attention and plaudits in the newspapers as unique events, all of them contributed to the overall American aims in the theater. At Black Rock, Elliott's cutting-out of *Detroit* and *Caledonia* had a threefold effect on the balance of power on Lake Erie. First, it reduced the Royal Navy's force by taking away its largest ship, which carried 30 percent of the Erie squadron's firepower, including half of their long guns, and removed the chance that the British would take a valuable smaller brig into service. Second, it offered an instant acceleration in the construction of the nonexistent American squadron by bringing *Caledonia* into the U.S. Navy and by providing guns for the armament of the private vessels, which shipwrights converted by

May 1813. Finally, *Caledonia* itself went on to important service with the squadron in Perry's victory in 1813, as well as in the Huron and Superior expeditions.[82]

At Cranberry Creek, American forces captured fifteen transports worth of naval materiel and added to American stockpiles at Sackets Harbor. The realization by Yeo and Prevost that they would need to take gunboat service more seriously also had a significant impact on the manpower in the region, leading them to take more than 250 troops away from their missions ashore to augment the undermanned naval forces. At Sandy Creek, the British defeat had a near-instant operational effect as the results forced Yeo to raise his blockade of Sackets Harbor because the loss of nearly two hundred men meant that he could no longer properly man his ships. Nearly the entire crews of both *Niagara* and *Montreal* were in the hands of the enemy, either prisoners or killed outright. Yeo's only available solution was to split the crews of *Netley* and *Magnet* just to have skeleton crews to sail the almost empty ships, never mind being able to fight.[83] Gregory's raids had an impact on the supply lines as well, raising alarms and taking British ships out of service by either sinking them, like *Black Snake*, or destroying them on the stocks as at Presqu'ile. With such small squadrons on Lake Ontario, every loss of a vessel could influence the balance of power. These results demonstrate how irregular operations can have just as important a strategic and operational effect in conflict as the loss of a ship in a broadside duel.

The actions on Lake Ontario also illustrate the role that aggressive junior officers and trusting senior officers play in raiding, reconnaissance, and cutting-out expeditions. Gregory, Elliott, Dimock, and Dixon all showed a spark of not only daring but also an understanding of how their exploits would contribute to the greater goals of their commanders. However, the capture of Gregory and the failures of the British units at Sandy Creek and Cranberry Creek demonstrate how the successful use of irregular operations tempers aggressive action with an understanding of the risks involved and the ability to balance between the two. Chauncey and Yeo both demonstrated how senior officers who understood the importance of irregular operations were vital to their successful use in a naval campaign. Chauncey's repeated use of Gregory and his raiders, as well as the praise he heaped on Elliott, illustrate his support for his lieutenants. British commanders on the lakes also appreciated the value of irregular warfare. Yeo himself attempted to lead a cutting-out expedition against the American ships at Sackets Harbor in late June 1813, but was foiled when deserters reported the scheme to Chauncey.[84] In 1814, Royal Navy commander Alexander Dobbs launched a successful cutting-out in nearly the same location where Elliott had made his captures, taking the American schooners *Ohio* and *Somers*, while *Porcupine* fought off the attackers and escaped.[85]

Finally, the operations conducted by both American and British forces on the lakes illustrate the importance of partnerships in the successful conduct of the missions. In some cases, like at Black Rock and Cranberry Creek, army or militia and naval forces operated successfully together. In the example of Gregory's raids and the American success at Sandy Creek, the local knowledge of, and intelligence from, lake mariners like William Vaughn and Samuel Dixon were important. Likewise, the contribution of allied Native Americans often played a central role. These kinds of local partners, or regional experts, are seen in examples throughout these chapters.

The history of raiding, reconnaissance, and cutting-out expeditions on Lake Ontario and the Niagara River during the War of 1812 reveals the role of naval irregular warfare as a part of a larger operational or strategic plan, shows the importance of aggressive but tempered junior leadership with supportive seniors, and emphasizes the role of partnership and cooperation. Yet the lakes were not the only theater that saw naval irregular warfare, and uniformed officers were not the only Americans who planned and executed irregular operations. Beginning in the spring of 1813, the efforts of civilian partisans took shape on the southern shores of New England and in the Chesapeake Bay. These irregular missions were triggered by the passage of the "Torpedo Act," and the promise of prize money for any American who sank a British warship, unleashing a combination of irregular warfare and innovative technology that would come to play an important role in the future of naval warfare.

Top
L. M. Quade, after Cecilia Beaux,
John Paul Jones, ca. 1956.
*Courtesy of the Naval History and
Heritage Command, NH 86493-KN.*

Botom
Ralph Earl, *Silas Talbot,* n.d.
*Courtesy of the Naval History and
Heritage Command, NH 85551-KN.*

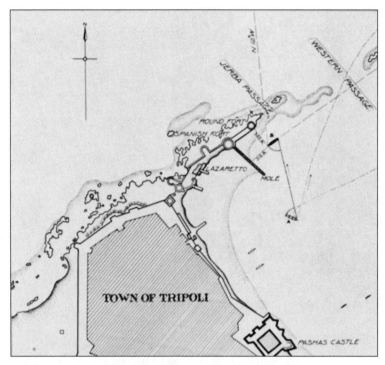

Chart of *Intrepid*'s mission to destroy the captured *Philadelphia*, 1804. This chart by
Charles Wellington Furlong depicts (by dot and dash lines) the course followed by
the ketch *Intrepid*, under Lt. Stephen Decatur, as she entered and left Tripoli harbor
on her mission to destroy the captured *Philadelphia*. Heavy dashed line indicates
Philadelphia's drift after she was set afire. Position A marks *Philadelphia*'s position
when boarded by Decatur's men, while B is the location of her wreck when Furlong
located it about a century later and recovered artifacts. Position X (just west of the
Jerba Passage) is the location of *Intrepid*'s wreck after she blew up on 4 September
1804, during another mission into Tripoli harbor. *Courtesy of the Naval History and
Heritage Command, NH 56745.*

Opposite, top
Experiment, 1799–1801. Eleven armed pirate boats attacked the *Experiment* on
January 1, 1800. *Courtesy of the Naval History and Heritage Command, NH 54116.*

Opposite, bottom
Edward Moran, *Burning of the Frigate Philadelphia in the Harbor of Tripoli,
February 16, 1804,* 1897. This painting shows *Philadelphia*, previously captured by
the Tripolitans, ablaze after she was boarded and set afire by a party from the ketch
Intrepid, led by Lt. Stephen Decatur. *Courtesy of the Naval History and Heritage Command,
KN-10849.*

Capt. David Porter, USN. Photograph by A. L. Brooks, of portrait, possibly by John Trumbull. *Courtesy of the Naval History and Heritage Command, 80-G-K-17588.*

Opposite, top

Rear Adm. Francis H. Gregory, USN, ca. 1861. Photograph by Matthew Brady. *Courtesy of the Naval History and Heritage Command, NH 49201.*

Opposite, bottom

Lt. Jesse D. Elliott, USN. *Courtesy of the Naval History and Heritage Command, NH 49625.*

U.S. schooner *Enterprise*, 1799–1823. Engraving by Jean-Jérôme Baugean, 1806. USS *Enterprise*, an American schooner with flags and sails out to dry. *Courtesy of the Naval History and Heritage Command, NH 621.*

U.S. schooner *Alligator*, 1821–23. Sketch of spars and sails by C. Ware, Boston Navy Yard, ca. 1840. *Courtesy of the Naval History and Heritage Command, NH 57010.*

Action of Quallah Battoo, Sumatra, as seen from the frigate *Potomac*, at anchor in the offing; J. Downes, Esq., Commander, February 5, 1832. Engraving, ca. 1830–39. *Courtesy of the Naval History and Heritage Command, NH 123364.*

Bombardment of Muckie. Engraving by Osborne, ca. 1840. *Columbia* and *John Adams* bombard the town of Muckie and land a force to burn the town on 1 January 1839. *Courtesy of the Naval History and Heritage Command, NH 1068.*

CHAPTER 5

Destructive Machines and Partisan Operations

The Torpedo Act and the War of 1812

The summer of 1813 brought a dramatic change in the Atlantic naval campaign of the War of 1812. In the first year of the conflict, the U.S. Navy strung together a series of victories at sea and early cruises that kept the United States in the war. Relying on these naval successes, Americans proudly held their heads high despite the repeated failures of the army in the attempts to invade Canada. However, 1 June 1813 marked a turning point, which shifted the focus of the war away from the blue water of the Atlantic and into the coastal waters, bays, and rivers of the Atlantic shoreline and the Great Lakes. It also marked the introduction of new tactics and weapons as both the U.S. Navy and the American people looked to fight the Royal Navy's ascendant forces. On 1 June two significant events occurred that changed the character of the naval war. First, off Boston harbor HMS *Shannon* patiently waited for the emergence of the 36-gun frigate *Chesapeake*. The American ship set sail under Capt. James Lawrence to respond to Captain Philip Broke's challenge for a frigate duel. The results of the battle have gone down in the history of both the United States and Royal Navies, as *Shannon* soundly defeated *Chesapeake* despite Lawrence's famous orders, "Don't give up the ship." The battle marked the end of a dramatic string of American frigate victories on the high seas.[1]

Farther south Commo. Stephen Decatur led his squadron, made up of the frigates *United States* and *Macedonian* and the sloop of war *Hornet*, east through Long Island Sound toward the open water of the Atlantic. Decatur's squadron sailed in an attempt to escape the British blockade, which was closing in around New York. In the vicinity of Montauk Point, at the eastern end of Long Island, the American lookouts spotted the approaching sails of the British blockading force. Outnumbered and outgunned, Decatur ordered his ships to turn and run. They set course for the harbor at New London, Connecticut, and there the Royal Navy blockaded the American squadron for the remainder of the war.[2]

The Royal Navy's success in the closing years of Britain's war with Napoleon freed ships from their wartime patrols in the Mediterranean, Baltic, and eastern Atlantic. In early 1813, the Admiralty sent many of those ships to reinforce Admiral Sir John Warren's squadrons, which were fighting the sideshow conflict with the Americans. Warren used these ships to strengthen the blockade of American harbors and trapped much of the U.S. Navy in ports up and down the eastern seaboard. With British ships off most of America's major cities, the naval war shifted into the green and brown water of the coastal bays, estuaries, and rivers and took on a new character.

In the Atlantic theater of the war, the efforts at coastal defense began to take on an irregular aspect. While the gunboats built during the Jefferson and Madison administrations had conceived of defending American harbors in conventional ways, with numerous small boats having battles and exchanging broadsides with larger enemy warships, these new operations involved stealth attacks and the initiation of undersea warfare. This included the introduction of new partisan fighters from the local watermen whose livelihoods suffered under the blockade, and the introduction of a new technology in naval warfare: the torpedo.[3] It introduced a naval conflict that focused on maritime security operations by the British to counter the naval irregular warfare adopted by the Americans.

Robert Fulton was already a famous inventor and engineer when, in the summer of 1810, he put his designs and theories for a new kind of naval warfare on paper in America. Raised during the American Revolution, he spent the years around the turn of the century in Europe, working first in Napoleonic France and then for Britain, designing revolutionary naval weapons. While he was in Europe, Fulton focused his design work on developing the infant field of undersea warfare. In France he designed and built *Nautilus*, which some historians have labelled the first practical submarine.[4] In 1804, Fulton moved to Britain where the government commissioned him to develop new weapons for the Royal Navy to use against Napoleon's invasion shipping. His work focused on the development of floating torpedoes, described today as sea mines, that defending forces could moor to the bottom, as well as methods for using the same explosive weapon offensively to attack enemy ships. After the victory of the British fleet at Trafalgar, however, the probability of invasion waned and along with it the need for elaborate coastal defenses.[5] In 1806, he returned to America and started his most famous work with his father-in-law, Walter Livingston: building the first commercial steamboat in the United States.[6]

In 1810, the possibility of a conflict between the United States and Great Britain was beginning to concern many Americans. The tension from the *Chesapeake–Leopard* affair had not subsided and British intransigence over the issues of neutral rights on the high seas and impressment, understandable based on their requirements in the Napoleonic war, increased the chance of war. In January of that year, Fulton demonstrated some of his designs to former President Thomas Jefferson and President James Madison.[7] Encouraged by their interest in his ideas he wrote a monograph entitled *Torpedo War and Submarine Explosions: The Liberty of the Seas Will be the Happiness of the Earth*, which was finished in the spring and sent to President Madison and the leadership of Congress. Fulton then arranged to publish it for the general public. In the pamphlet Fulton described in detail his designs for underwater explosive devices that he called "torpedoes" and the tactics that could be used to employ them against an enemy's ships.[8]

Drawing on the results of his designs and experiments in Britain, which included the sinking of the test ship *Dorothea* in the Downs off the coast of Deal at Kent, as well as less successful testing in combat during the Royal Navy's raid on the French port of Boulogne, Fulton outlined a system of harbor and coastal defense. He maintained that it would revolutionize naval warfare. In the attack on *Dorothea*, Fulton proved a canister of powder sunk below the hull of a ship and detonated could sink the vessel. From that starting point, he developed three primary tactics for deploying the weapons. First, he created torpedoes anchored to the bottom, suspended beneath the surface in harbor entrances and detonated by contact fuses. Second, he designed floating torpedoes deployed from boats to take advantage of current and tides to strike enemy ships at anchor or during their approach to a harbor. Finally, he worked on a scheme of torpedo boats, which would attack in swarming groups. These boats could either use long spars projecting from their bows to attach the torpedoes to the hull of an enemy ship or, in his more dramatic suggestion, fire harpoons from guns mounted on the boat, which would attach the torpedo while the boat's crew rowed away to safety.[9]

The pamphlet was convincing and on 30 March 1810, Congress passed "An Act Making Appropriation for the Purpose of Trying the Practical use of the Torpedo, or Sub-marine explosion," which designated $5,000 to fund experiments with Fulton's system.[10] Secretary of the Navy Paul Hamilton, in consultation with leaders in Congress, ordered that the test be made on the Potomac River to allow observation by interested parties. In April, Fulton wrote to him and suggested a preliminary set of tests conducted in New York harbor. Hamilton agreed to the New York tests as long as Fulton committed to follow-on tests on the Potomac, and disbursed $1,000 to fund the work in New York.[11] The secretary appointed a panel of judges to assess the initial tests, and in the summer, the panel met in New York City to hear Fulton's ideas and watch a demonstration. Secretary Hamilton assigned Commo. John Rodgers and Capt. Isaac Chauncey from the New York Navy Yard in Brooklyn to assist the panel.[12]

Fulton described his weapons to the panel and outlined a number of exercises that he intended to use to demonstrate their effectiveness. He suggested, in his pamphlet, that his system had the potential to obviate the need for a navy.[13] It comes as no surprise that Commodore Rodgers felt such an assertion was foolish, at best; excerpts from his journal reveal an obvious defensiveness. Fulton asked to use the frigate *President* as his test ship, intending to attack with torpedo canisters emptied of their powder. Since *President* was in the North River undergoing repairs to its rigging and scraping of the hull, Rodgers instead suggested the sloop *Argus* that was in the East River, anchored off the Navy Yard. The captain ordered James Lawrence, the lieutenant commanding *Argus*, to prepare her for the attack and showed him how to rig the ship to create barriers around her, using a splinter net and the studding sail booms anchored by pigs of ballast iron, which the torpedoes would have to penetrate to get beneath the hull.[14]

When he saw the defenses constructed around *Argus*, Fulton admitted his current system was unlikely to penetrate the perimeter created by the net on the bow and the booms floating around the sides. He demonstrated his harpoon gun, designed to attach torpedoes from a distance, but was only able to attach the harpoon to a hull at 15 feet, inside of where the booms lay in the water. For each idea he presented to the panel for how to get the torpedoes past the defenses, Rodgers and Chauncey had plans for how they could adjust their efforts. The back-and-forth continued in front of the panel until Fulton gave up, admitting that he had more work to do.[15] In the panel's report to Secretary Hamilton, they predictably stated that they did not believe Fulton's system could replace a navy in defending America's shores. However, they also recognized the suggestion was a high bar to cross. As the early stages of the coming war would show, the U.S. Navy itself also struggled to defend the American coast. The panel stated, "[W]e, however, disclaim the intention of attempting to discourage such investigations and experiments." In their opinion, Fulton suffered from a lack of practical experience experimenting with his system. It showed promise and they recommended that the navy and Congress continue to fund experiments to develop the theories and weapons.[16]

In his after-action letter to Secretary Hamilton, Fulton admitted he needed more time to experiment with his designs and to develop and practice the tactics and techniques needed for successful employment of the weapons. He wrote: "I unite with the committee in the opinion that the Government should not rely on this, or any new invention, for the defence until its utility be fully proved. It never has been my wish that such confidence should be placed in torpedoes, until fair experiment had proved their value beyond a doubt."[17] In the letter, he discussed the improvements to the system he had already developed from the lessons of the experiments and stated his continued belief that the innovative new weapons would become critical to the future of naval warfare. Torpedo warfare had its first demonstration and discussion, and conservative leaders of the navy Commodore Rodgers and Commodore Chauncey had managed to keep the new and different at bay.[18]

The dismal showing in the New York tests was not enough to shake Fulton's confidence that he was developing a weapon with revolutionary potential. Following his February 1811 letter to Hamilton describing the results of the tests, he wrote again in April to explain he had more testing to conduct and to request more funds. Hamilton responded by again disbursing $1,000 so Fulton could continue his experiments.[19] A year later, just a month before the declaration of war, Hamilton ordered release of another $500 for Fulton to continue work, but warned him that his experiments must be completed by the end of 1812 since the original appropriation would expire.[20]

On 18 June 1812, the United States declared war on Great Britain. Fulton saw his opportunity and, less than a week after the declaration of war, he sent a letter to Hamilton. The inventor had placed most of his torpedo equipment, test units, and "various apparatuses" in storage at the Washington Navy Yard. He asked that Secretary Hamilton have them shipped north to New York because "before the termination of the present War we may expect a visit from British ships of war in this port; and I should like to be prepared for such an event in the best manner I can." Fulton also asked for the balance of the $2,500 from the congressional appropriation to fund his operations out of New York. Secretary Hamilton, however, was busy with other issues related to the U.S. Navy's preparedness for war, or lack thereof, and the letter remained unanswered.[21]

―――――――――

In the spring of 1813, the American war effort was riding a high tide from the frigate victories of the previous year. However, that tide of war was turning. With new ships sent to North America Station, the Admiralty ordered Admiral Sir John Warren to tighten the blockade of American ports. The Admiralty's new strategy was to clamp down on the American economy by "placing all of the enemy's ports in a state of close and permanent blockade," while ensuring the safety of British commerce and war shipping by managing a well-protected convoy system.[22]

It was apparent to Americans in the early months of the new year that the blockade that the Royal Navy had announced in late 1812 was not just going to be a paper blockade. Efforts by the U.S. Navy to defend American ports were met by competing priorities for the Navy Department. Secretary of the Navy William Jones took over from Hamilton in January of 1813 and was in desperate need of men and materiel to construct and equip the squadrons the department was building on the Great Lakes. Knowing that the navy's gunboat flotillas were in a state of disrepair, he ordered a reduction in the number of active gunboats, instructing the station commanders to use the opportunity to lay up their weaker boats, strengthen the boats they kept in service, and provide the leftover men to the department for reassignment.[23]

While the policies of the Navy Department might have been wise, considering its limited manpower and budget, there was still a need to defend American harbors. On 13 March 1813, Congress passed "An Act to encourage the destruction of the armed vessels of war of the Enemy."[24] The new law stated that not only was it lawful for any private citizen to attack British warships, but the Treasury of the United States would pay a bounty of one-half the value of the vessel if it was burned, sunk, or destroyed, as well as one-half of the value of the cannon, cargo, and equipment on board. Congress suggested "for that purpose to use torpedoes, submarine instruments, or any other destructive machine whatever."[25] Based, it appears, on the well-accepted use of privateers mixed with inspiration from Robert Fulton's innovative weapons, the law became known as "The Torpedo Act." In most respects, it echoed Fulton's letter to Secretary Hamilton at the start of the war, in which he suggested that the government offer a bounty on enemy warships to encourage attacks.[26]

Word spread of the passage of the Torpedo Act and private citizens set themselves in motion. Fulton, still living in New York, wrote to President Madison: "The law of Congress in favour of Torpedoes and Submarine instrumts [sic] will I believe be productive of happy consequences," and announced he was drawing up plans for attacks.[27] He planned to work with others who might sponsor or fund the operations, including Governor Morgan Lewis of New York, who had been one of the panel members in 1810 and one of Fulton's supporters in the group.[28]

Fulton's pamphlet on the uses of torpedoes and how to deploy them was still available in New York in the spring of 1813. However, the inventor had written it before the New York harbor tests, which had gone so poorly. In the weeks following the passage of the Torpedo Act, Fulton set to work on a document that would serve as an instruction manual for persons "willing to try their fortunes," as Master Commandant Jacob Lewis would eventually describe them.[29] The design for the actual weapons was something Fulton kept relatively secret, so his manual focused on the tactics that attackers might use. Drawing from more than a decade of developmental tests, including those made with the Royal Navy, the operational testing during the raid on Boulogne, and the efforts with the U.S. Navy, Fulton updated his methods. He suggested techniques of intelligence collection, advantageous use of weather conditions, and the ways the weapons could be deployed. Fulton laid out what today's naval officers and sailors call a manual for tactics, techniques, and procedures.[30]

In April, Fulton wrote to Thomas Jefferson and assured him: "I am not Idle as to Torpedoes, but secrecy is necessary."[31] He was in the process, with his friend and occasional business partner Benjamin Henry Latrobe, of recruiting a mariner named Elijah Mix for a torpedo expedition against the Royal Navy in the Chesapeake Bay. Immediately following the passage of the Torpedo Act, Mix had put a plan in motion in Baltimore. Master Commandant Charles Gordon, commander of the naval forces in Baltimore in the spring of 1813, wrote to Secretary Jones about Mix's proposal to the city's Committee for Defense during the second week of March. He planned to put together a number of torpedo boats, all based from a larger ship, which the city would provide. The navy had authorized Gordon to lease a small force of former privateers to serve in the Chesapeake.[32] With "10 or 12 of [Mix's] boat torpedoes in the different vessels," Gordon saw the chance for an innovative new unit that he could add to his squadron. Despite his apparent approval of the idea, Gordon left it up to the Baltimore committee to support Mix and his scheme.[33]

Less than a month later Mix and Fulton had an arrangement on paper. The deal, coordinated by Latrobe, was that he and Fulton would provide the torpedoes to Mix in exchange for a third of the bounty money paid out upon success.[34] That same week Fulton first wrote to Secretary Jones in an attempt to determine the new secretary's attitude toward his innovative mode of naval warfare. Fulton laid out a brief case for the torpedo and assured the secretary that any endeavor would have his full support.[35] While Jones does not appear to have responded directly, the secretary's actions suggested limited but official support for torpedo operations. Less than a month later Fulton submitted an application to patent "several improvements in the art of Maritime War," which included the torpedo. He went so far as to send a letter to Secretary of State James Monroe, via Latrobe, looking for support in his claim for the patent.[36]

Prior to Latrobe and Fulton's correspondence about partnering with Elijah Mix, Mix had written directly to President Madison with a plan for a torpedo attack.[37] He sent a more detailed plan in the middle of the month, after consulting with Secretary Jones. He partnered with Frederick Weedon, a Chesapeake Bay man from farther south in Mathews County, Virginia. Weedon owned property on the water, and a British 74-gun ship of the line regularly lay within sight of his waterfront just north of Mobjack Bay.[38]

On 7 May, Secretary Jones sent a confidential order to Charles Gordon to provide Mix with 500 pounds of powder, a boat, and six volunteers if he could find them, for the attack.[39] Three days later Gordon reported to Jones that Mix had spent the day with him, and he would have a completed torpedo ready the next evening. The next step was to mount it on a vessel, which the Baltimore Committee had still not agreed to provide. Mix and Gordon were working to obtain "an old brig" that could serve to carry the torpedoes and boats closer to their targets, and Gordon volunteered to help victual and man the brig in order to have it sail with his schooners when they departed on a reconnaissance of the bay. He wanted to ensure that they would have the new weapons available if his ships found an opportunity to employ them.[40]

The effort to obtain the brig fell through and Gordon reported on 19 May that Mix and six men in a fast rowing boat had departed Baltimore and headed into the bay to join him on board *Revenge*. Gordon sailed for the southern part of the Chesapeake the next day for a month-long scouting cruise.[41] After exercising his ships near Annapolis, *Revenge*, the schooners *Comet*, *Patapsco*, and *Wasp*, and *Gunboat No. 138* cruised the middle to southern portion of the Chesapeake, observing the British squadron. The small squadron's operations included a number of visits to New Point Comfort, at the southern tip of Mathews County, Virginia, and may have delivered Mix, his boat, and crew of volunteers.[42] At some point, however, the two expeditions parted ways because on 6 June, from off the mouth of the Potomac River, Gordon reported to Jones with some measure of disappointment, "I have never yet seen or heard of him."[43]

The night before Gordon sent his report, HMS *Victorious* was in the southern Chesapeake Bay and discovered an object drifting in the water. The crew recovered it and determined it was "one of the powder machines, commonly by the name of Fulton's." Captain John Talbot described it as a raft made up of six barrels of powder, which were counterweighted to lay just below the surface, with lines running out of each side and connected to buoys. The British estimated in a close anchorage, where the lines to the buoys would likely wrap around a ship's anchor cable and draw the device in close to the hull, it would likely have been very dangerous. It was lucky they recovered it in open water.[44] Rear Admiral Sir George Cockburn, in command of the Royal Navy's squadron in the Chesapeake, reported the event to Warren. He explained he had "closed with His Majesty's Ships toward Hampton Roads" because of the attack and he believed that by concentrating his defenses any future enemy attempts "will prove equally futile and unavailing with respect to their Effects [*sic*] upon us."[45]

If Cockburn, as he said in his letter, pulled the squadron back toward Hampton Roads because of the torpedo, it is probable *Victorious* was north of the York River at the southern end of the Chesapeake when it discovered the "Fulton." This would have placed the ship off of Mathews County, Virginia, where Mix had planned to meet up with Weedon according to the plans he described in his letter to the president. The amount of powder that would fit in six barrels is close to what Gordon and the navy provided Mix in Baltimore, and the description of the device in British sources indicates it was a primitive design when compared to Fulton's copper-cased weapons. All of these details match up with the kind of device Elijah Mix, who was neither an engineer nor a weapons expert, had built himself in Baltimore. This evidence suggests that, despite what other historians have written, the waterborne improvised explosive device recovered by *Victorious* was not a random encounter, but instead a failed attack by Mix and Weedon and the first torpedo operation of the war.[46]

The first fully documented attack on the Royal Navy following the Torpedo Act was in late June, following the appearance of the newly reinforced blockading squadron in Long Island Sound and Decatur's retreat to New London. A New York City merchant named John Scudder led a group that rigged a local schooner to carry out an attack on the blockading squadron. With the assistance of Master Commandant Jacob Lewis, commander of the U.S. Navy's New York gunboat flotilla, Scudder and his compatriots loaded the bottom of the vessel's hold with 400 pounds of powder mixed with sulfur and surrounded by stones and other items meant to serve as shrapnel. Above the explosives, they loaded the rest of the hold with barrels of flour, peas, and naval stores that they knew the British needed. They attached a pair of flintlocks to the barrels of powder, with trigger mechanisms initiated by a line to the bottoms of the barrels of flour. When a prize crew removed the stores, the explosive device would detonate. Lewis suggested the group select a vessel with a main mast that was about the right size to make a good topmast for one of the British ships of the line, and that they rig the explosives to the step of the mast. In order to remove the mast, the British would be forced to bring the vessel alongside one of their larger ships, ensuring a valuable target. Finding such a vessel was determined to be prohibitively expensive and the group settled on the small schooner and the arrangement of the stores.[47] On the morning of 25 June 1813, *Eagle* sailed from New York for New London, where HMS *Ramillies* was anchored watching the American squadron.[48]

The British blockading squadrons did not receive regular resupply and, for the most part, they were dependent on captured American coasters or raiding ashore to resupply their ships. At ten o'clock, the British spotted the schooner sailing toward New London. Captain Thomas Masterman Hardy, commanding HMS *Ramillies*, dispatched a master's mate with a small crew in one of his ship's boats to intercept the American vessel. The Americans saw him coming, and with light winds they had time to get *Eagle* in toward shore where they abandoned the vessel. While they appeared to "escape" from the British attackers, they dropped the schooner's anchor and made sure it held, so she would not ground and make the British capture more difficult. On shore, the Americans fired a "sharp musketry" to harass the British, and make the capture look contested, but Master's Mate William McIntyre and his crew took possession of *Eagle* by eleven o'clock.[49]

By one o'clock, the Royal Navy crew had sailed *Eagle* back to *Ramillies*. However, McIntyre cut the cable rather than recover the anchor because the Americans had him under fire. Captain Hardy sent Lieutenant John Geddes with *Ramillies*'s pinnace and a small crew to help McIntyre moor the schooner to an American sloop they had captured the previous day. Once alongside, the sailors began furling the sails and lashing the vessels together. Hardy stood on his quarterdeck and watched what appeared to be an efficient evolution when "the schooner blew up with a tremendous explosion, and I lament to say Lieutenant Geddes and ten valuable seamen fell a sacrifice to this new mode of warfare."[50]

Hardy admitted in his report to Warren that it was simple luck they had not brought *Eagle* alongside *Ramillies*. Jacob Lewis chalked it up to the tides and winds making mooring to the captured sloop easier, but he lamented the fact that the attackers had not followed his plan of selecting a larger vessel and rigging the explosives to the step of the mast.[51] Hardy moved his ship's anchorage farther away from New London, closer to Long Island. In reporting the movement to Secretary Jones, Decatur predicted that if things had gone according to plan "the explosion would have destroyed the ship."[52] Recognizing the luck that *Ramillies* experienced, and the danger posed by such attacks, Admiral Warren issued General Order Number 87, instructing captains not to bring prizes directly alongside His Majesty's ships and vessels, or allow American boats to approach too close. After a proper inspection had taken place, crews could bring the prizes and boats alongside.[53]

Less than a week later, a second attempt was made on *Ramillies*. In the spring of 1813, Silas Halsey of Norwich, Connecticut, built a "diving boat" along similar lines to the submarine device built by inventor David Bushnell during the American Revolution. Halsey designed his boat to dive under the hull of an enemy ship, where he would use a drill bit at the end of the propeller to put a hole in the bottom and attach a torpedo. On the evening of 30 June, Halsey set out from New London in his boat to strike at Captain Hardy and his crew. Accounts differ as to how far he was able to get. Decatur, in his report on the event to Secretary Jones, stated that he was not gone from the harbor much more than five minutes when British sympathizers on shore signaled to *Ramillies* that he was coming and he was unable to complete the attack.[54] A later account from a British source states the submarine surfaced "like a porpoise for air" and it was then that lookouts spotted it and Hardy ordered the cable cut and the ship got under way. However, it appears Halsey may have attempted to strike twice more. The movement of the ship thwarted his second attack and on the third attempt, he came up directly under *Ramillies* but had his drill bit break off in the process of securing the torpedo to her hull.[55]

Fulton himself doubted the attack actually took place, instead believing it was an elaborate scare tactic employed against the British. He asked Decatur of Halsey and the diving boat: "[D]id you even see him in it?"[56] Regardless, the results of the attack were clear. Halsey did no material damage to *Ramillies*, but the threat was enough that Hardy left his anchorage and kept his ship under way. He then moved his anchorage across the sound toward the end of Long Island, at Gardiner's Island, where he perceived the shoreline to be safer, supplies easier to come by, and his blockade still fully enforced.[57]

Back in the Chesapeake, following Gordon's return to Baltimore, Elijah Mix, now officially warranted as a sailing master in the U.S. Navy, and his crew remained behind in the southern part of the bay.[58] The British squadron stretched from Hampton Roads across to Lynnhaven Bay at the far south of the Chesapeake. A few ships were in the vicinity of Cape Charles at the bay's eastern limit where it opens to the Atlantic. The ship farthest east was HMS *Plantagenet*, a 74-gun ship of the line that remained anchored near the Cape Charles light. Based on its position farthest from the bulk of the fleet, Mix decided to target it. On 18 July, he took his boat, which he named *Chesapeake's Revenge*, to make an attack with two torpedoes he obtained from Robert Fulton in May.[59] Interrupted by one of *Plantagenet's* guard boats while trying to deploy his torpedo, Mix and his crew recovered the weapon and rowed away safely before the British could come alongside.[60]

Over the course of the next four nights Mix and his crew, including a Massachusetts waterman named Bowman and Midshipman James McGowan, made several more attempts on *Plantagenet*. On 19 July, they had little luck, but on 20 July, the party closed beneath the jib boom of the ship and was readying the device when a lookout spotted them and fired a musket in alarm. Mix and his men pulled hard at their oars, while the British put boats in the water and commenced firing small arms in their direction. Losing the Americans in the night, the British fired rockets over the water to illuminate the darkness and allow them to pick out the silhouette of the fleeing Americans to direct their fire. Once again, however, the Americans escaped. *Plantagenet* began to shift her position each evening from 20 July to 23 July; the ship and the American crew played a game of cat and mouse as each night the Americans returned to the vicinity, but each night the British had changed their anchorage.[61]

On the night of 24 July, Sailing Master Mix and his party found *Plantagenet*. Bringing their boat within one hundred yards of the ship's port bow, they deployed their torpedoes and watched as the tide began to carry them toward the ship. Unfortunately for Mix and his men, the weapons detonated early (though the reason for the early detonation was apparent to them, according to reports). Captain Robert Lloyd reported that a massive explosion occurred about half a cable length from his ship, throwing a column of flame and water into the air.[62] At first, it alarmed the British crew, and Mix reported to the *Norfolk Herald* that he and his men saw the British rushing for their boats to abandon ship.[63] While it is true that they were rushing to their boats, reports of prisoners on board *Plantagenet* later stated it was not to abandon ship but instead to set off in pursuit of the torpedo party.[64] Mix and his men escaped into the darkness, and once ashore he began working on another set of plans.

The Delaware Bay was not quiet either. A group led by Samuel Swarthwout devised a plan to use a pair of boats and floating torpedoes with contact detonators to make an attack against the British, who had been raiding in the bay. In April, the two boats and twenty men set out, eventually locating the British 74-gun HMS *Poictiers*. However, as they prepared for their attack Governor Joseph Haslet ordered a halt to the operation, because the British were holding American prisoners on the ship. The operation was abandoned, but in June the boats were back on the water in search of a target. On 30 July, they sortied against HMS *Statira* but without success.[65]

The torpedo attacks against his ship incensed Captain Hardy on board *Ramillies*. Following the *Eagle* attack and Halsey's diving boat, Hardy wrote to New London under a flag of truce, intent on determining if the attacks had been carried out by the government or by private citizens. If they were government-sponsored attacks, he threatened he would "destroy everything American that floats." The local leaders in New London assured him that members of the U.S. Navy did not conduct the attacks, an assurance that, while accurate in letter, was questionable in spirit. He tightened his blockade, and his fellow officers began warning neutral ships of the hazards of the "inhuman and savage proceedings" and that the Americans may use them to attack them as well.[66]

In August, Hardy became aware of a torpedo plot to his south along the shores of Long Island. He had made Gardiner's Bay the central rendezvous for the blockading squadron. *Ramillies* and the other ships frequently spent time in the anchorage there, across the sound from New London. His network told him that in the web of narrow, protected inlets just off the southern and eastern edges of Gardiner's Bay, a plot was in motion. A local man named Thomas Welling was preparing a torpedo attack with a number of his neighbors using a boat on his property, including a man named Joshua Penny.[67] During the last weekend in August, Hardy sent a landing party of Royal Marines ashore, commanded by a lieutenant and under the cover of night, to disrupt the attack. Going directly to Penny's home they took him out of his bed, shackled him while he was still wearing his bedclothes, and returned to *Ramillies* with him and a neighbor named Robert Gray. Penny later wrote that the lieutenant told him he had orders to burn his house to the ground, but because his wife and children were present he was merciful and left it intact.[68]

Once they had Penny on board, Hardy's officers recognized him. He had been on board *Ramillies* several times to sell clams and fruit to the British, and had been collecting information. Hardy also learned that Penny had served as a pilot for a small-boat expedition that Decatur's men had launched out of New London earlier in the summer, and the Americans had paid him for the service. Maj. Benjamin Case, commanding the U.S. troops at Sag Harbor, sent a request to Hardy from the residents of East Hampton asking that he release Penny because he was a civilian.[69] Claiming his work as a pilot meant he was "on the books" of Decatur's squadron and a legitimate prisoner-of-war, Hardy refused and eventually took Penny to Halifax. Determining that Robert Gray was "an inoffensive old man," Hardy did agree to release him.[70]

Although the British had disrupted the plot, Hardy's men had not discovered the torpedoes or the whaleboat allegedly rigged for their deployment. The day after he took Penny, Hardy sent a letter to the justice of the peace of Southold in East Hampton, New York. He told the residents of Southold he knew there was a torpedo plot under way, that he had very good intelligence, and that he was going to continue to pursue Thomas Welling and the men of "mercenary motives" who were after his ship. He closed the letter by warning the inhabitants "of the towns along the coast of Long Island, wherever I hear this boat, or any other of her description has been allowed to remain after this day, I will order every house near the shore to be destroyed."[71] Whether because of Hardy's threats, or other reasons, the Welling boat never made an attempt on *Ramillies*.

By the end of August 1813, Fulton's thinking about torpedo boats and their methods had begun to shift. He was beginning question whether small boats with a handful of men, while effective and efficient in small bodies of water or protected areas like harbors, were operationally viable in larger bodies of water like Long Island Sound or the Chesapeake. He suspected slightly larger craft might be more effective. His design work led him to think about larger craft for torpedo operations. He focused on vessels the size of a small schooner. On 5 August, in a letter to Decatur, Fulton first raised the idea of a steam-powered torpedo boat. Built around his previous submarine design, he laid out the first concepts, which would eventually become *Demologos*, the U.S. Navy's first steam-powered platform.[72]

Less than a month later, however, Fulton was back to designs that were more conventional. Still focused on a larger vessel, better for keeping at sea than a six-man whaleboat, Fulton settled on a fast-sailing schooner with a 90-foot length and a 23-foot beam, along the lines of a design built by the New York naval architects Noah and Adam Brown. He suggested using pine in the construction because, while it was normally less desirable in ship construction, it was inexpensive and was likely to provide decent protection against musket fire. He envisioned building thirty-three schooners and fitting each with an underwater spar that extended beyond the bow. The attackers would attach a torpedo with a contact detonator at the end of the spar. The length of the spar, mounted to the hull with thirty feet extending beyond the bow, was likely derived from the protection provided by the barriers and nets, which the British had adopted, similar to how Rodgers and Lawrence defended *Argus* in 1810. Fulton sketched out the tactical employment of the "fast schooners," showing a swarm attack with fifteen torpedo boats against a single British ship of the line. The boats would approach in a bow-on attack stretching from the port to starboard beam.[73]

In their early August correspondence, Fulton and Decatur discussed a pair of plots they believed were under way. The first was planning for another attack by John Scudder and the captain who piloted *Eagle* while luring in the British in June. The second was by a pair of men Decatur had learned about, described as "a Major Fink & Mr. Richards." However, despite the interest of Fulton and Decatur and their intent to assist the two efforts, neither appears to have amounted to an actual attack.[74]

In the closing days of June, Fulton sent an associate, a fellow New Yorker named James Welden, to New London with "a plunging boat[,] torpedoes and apparatus" and $200 in advance for expenses. Welden and Fulton had signed a contract to split the proceeds from a torpedo attack, with the equipment and expense money provided by Fulton and the work completed by Welden. If the man did not survive the attack, the contract was clear that his heirs or the executor of his will could designate where his half went.[75] In a 9 August letter to Fulton, Decatur told him "your man Welden is here" and that the full moon would likely delay any attack. Decatur approved of the man, saying he "appears prudent and perceiving," and stated that he would help him in any way he was able. He closed the letter by saying, "I shall offer more remarks on the submarine business shortly."[76]

On 24 August, a night before the new moon, a torpedo boat attempted an attack against the British. The New York City newspaper *The War* reported, "[I]t is said the torpedo from New York was chased on Tuesday of last week nine miles, by several British boats, but by repeated diving escaped." The newspaper stated that the British deployed extra guard boats to ensure the attacker could not make another attempt.[77] Coming late in the month, near the time of the new moon as advised by Decatur, it is likely that this was Welden's boat, and another Fulton-inspired attack was thwarted. It was the last attack of 1813, as the cold of fall and winter set in, making life on the water less hospitable to the raiders.

The summer of 1814 approached and brought more seasonable weather for partisan attacks, and American torpedomen began plotting again. On the night of 24 March, a boat closed on HMS *La Hogue*, captained by Thomas Capel and serving as the command ship in Long Island Sound while *Ramillies* and Hardy were at Halifax for refit and resupply. Towing a 30-foot-long tin tube, seven inches in diameter and full of powder, Capt. Jeremiah Holmes commanded a small party of men in the attack. Like Joshua Penny, who was involved in operations the previous summer, Holmes had been impressed into the Royal Navy and escaped prior to the war. Torpedo operations offered a way to exact his revenge. While the design of the weapon was different from Fulton's copper torpedoes, Holmes followed a tactical plan similar to how Mix had used the weapons Fulton provided him on the Chesapeake. Towing the torpedo into place and then allowing it to drift down on the target, Holmes launched his weapon in the night. But, also like Mix, the mission met with failure. Rather than impacting the hull of *La Hogue* and detonating, the long torpedo became entangled with the warship's anchor cable and exploded prior to reaching the ship itself. With a dazzling display, and water shooting into the air, it made a great deal of noise but had little tactical effect.[78]

As the third year of the war continued, and the blockade choked American harbors, the Common Council of New York City decided to finance an attack in defense of their harbor and merchant trade. Working from a design by a New Yorker named Berrian, they funded several hundred dollars for the construction of a new kind of torpedo boat.[79] This vessel was a semi-submersible: about 30 feet long with a cabin top just a foot above the surface of the water and armored, "on her back a coat of mail." The craft had a crew of nine and was able to deploy three torpedoes from a spar on one end of the craft. Described in the newspapers as "resembling a turtle" the craft was completed "at great expense" and left New York at the end of June.[80]

Sailing east along the north shore of Long Island, a pair of gunboats accompanied the turtle boat until it was in the vicinity of Southold, a few miles from the British anchorage at Gardiner's Bay. The gunboats departed, and the crew anchored near Horton's Point to observe the British and identify their target. Unfortunately, Horton's Point does not provide much shelter from weather or seas, and when the boat anchored the weather began to deteriorate. The seas built and the boat began to pitch at anchor. One member of the crew panicked, climbed through the hatch, and jumped overboard to swim to shore. He struggled to keep his head above the rising waves, and the crew cut the anchor to attempt to drift down on him and pull him out. However, the seas put the boat hard aground on Horton's Point. The crew was able to make their escape, but the British blockading force sent the frigate HMS *Maidstone* and sloop *Sylph* to reconnoiter the wreckage. They landed on the beach next to the remains and had a brief firefight with the Sag Harbor militia. The historical sources are conflicted over whether it was the British who destroyed the boat or the crew who blew it up in order to keep the British from capturing it, but nothing was left of the innovative craft.[81]

Long Island was not the only scene of 1814 efforts. In the British attack on Washington, D.C., in late summer, there are references to torpedoes in the letters and reports detailing the short-lived invasion. Following the burning of the White House and other buildings, as the British ships withdrew south down the Potomac River, Capt. David Porter sent a torpedo boat after them. He reported to Secretary Jones that he heard an explosion from the direction of the enemy squadron but did not have an immediate report of the result.[82] Fulton himself was in Washington in the days leading up to the British assault. He recovered some of his torpedoes from the Navy Yard and took them to Baltimore to assist in the city's defense. However, when he arrived he discovered Commo. John Rodgers, who had helped quash the 1810 experiments, in command.[83] Fulton, apparently willing to set aside their past conflict over undersea warfare, visited Rodgers's home and showed him some models and some of his detonators before returning to New York. When the British force descended on Baltimore harbor, Fulton sent Rodgers a letter highlighting the potential of his invention to help the defenses.[84] But Rodgers ignored him again. Despite the fact that the situation at Baltimore had a combination of geography, currents, and enemy disposition that made it ripe to become the first large-scale torpedo success, Fulton's torpedoes were not employed.[85]

The final plot for a torpedo attack during the war occurred late in 1814 on Lake Ontario. Midn. James McGowan, who had served with Elijah Mix as a volunteer on the Chesapeake the year before, led an expedition from Sackets Harbor in a whaleboat with ten men and a number of torpedoes. Their mission, as ordered by Commodore Chauncey, was to slip up into the St. Lawrence and ascend to Kingston from below. Once in the harbor they were to deploy their torpedoes with the goal of destroying the British ship of the line that was under construction and being fitted out.[86]

The crew left Sackets Harbor after sundown on 12 November 1814, with nine sailors and a local man named Johnson, who was the pilot of the new American frigate *Mohawk*. They reached the shore opposite Fox Island, about half the distance to Kingston, before midnight, pulled their boat ashore and set up camp. Squalls and rain built through the night and throughout the day on 13 November, with McGowan reporting "fresh gales and heavy rain" that lasted all day. He kept his men busy preparing for their mission by readying the torpedoes and equipment despite the cold rain. On the afternoon of 16 November, the expedition repositioned to the mouth of the St. Lawrence, where they waited for dark, and then proceeded down the river for seven miles before making camp again. After spending the day hidden on the American side of the St. Lawrence, the expedition moved downriver again in the dark, pulling their boat out of the water at Mill Creek. The moon was bright, and McGowan elected to remain ashore until his crew had a dark night to round the end of Long Island (known in Canada as Wolfe Island) and head back up river to Kingston to make their attack.[87]

The next day the weather worsened further and the crew remained encamped through the night of 16 November. As they waited for a break in the weather, the Americans spotted a pair of boats rounding the end of Long Island and heading for the American shore. They appeared "armed and well manned," and McGowan ordered his men to "make every necessary preparation for receiving them." Waiting patiently for the boats to approach the shore, after an hour and a half the Americans rushed out of their hiding place and caught the British completely by surprise, demanding their surrender. Whether afraid there were more Americans on the beach, or hiding in the trees, or simply unprepared to fight in such terrible weather, the British surrendered.[88]

But McGowan found himself in a difficult position. His mission was to attack the ships fitting out at Kingston, but he now had as many prisoners as he had men in his expedition. After securing the prisoners and their equipment, he decided to forgo the torpedo attack and return to Sackets Harbor with his prizes. Waiting until the next evening to slip out of the St. Lawrence under the cover of darkness, he left behind a line of torpedoes to slow the pursuit of the British gunboats he knew were patrolling the area. The three boats arrived back in Sackets Harbor on the morning of 19 November, after eight days of hiding during the day and moving through the cold November nights. McGowan stood by his decision, and the captured boats and prisoners brought good intelligence on the British squadron's movements and construction in Kingston. Commodore Chauncey approved as well. While he considered sending another torpedo expedition "before the frost sets in," the winter approached too quickly for McGowan to make a further attempt.[89]

Most histories of the torpedo operations during the War of 1812 have cast these attacks as purely the work of civilians, or simply related them in a regional context. Seen as something akin to privateers, these partisans are commonly viewed either as dedicated patriots or as profiteering scoundrels.[90] However, in reality, the work of these parties was anything but independent. Despite the centrality of a technological change, they displayed clear similarities with other examples of naval irregular warfare in the Age of Sail. The people involved continued to demonstrate the importance of aggressive young officers and open-minded seniors who were willing to understand and balance irregular risk. The role of innovative technology in this example sets it apart from the other episodes studied here, but at the same time, it continues to highlight the role of small combatants and the need for appropriate platforms in irregular operations. The relationships between the operators in the form of civilian partisans, naval officers who provided logistical and occasionally operational support, and the nexus point provided by Robert Fulton, demonstrates the continued relevance of partnerships in irregular warfare.

Throughout history, some naval officers have been conservative and reluctant when it comes to radical changes in methods and tactics. The introduction of torpedo warfare was no different in that regard to the development of steam warships or the introduction of naval aviation.[91] Senior officers during the War of 1812 can generally be classified in two groups: those who conducted maritime security and irregular operations as junior officers during the Quasi-War with France or the Barbary War, and those whose formative naval experiences focused on conventional blue-water, ship-on-ship engagements and missions.

In the example of War of 1812 torpedo operations, the two archetypes of senior officers are demonstrated by Commo. John Rodgers on the one hand, and Stephen Decatur on the other. As Robert Allison wrote, "Rodgers had regarded Fulton as a madman; Decatur thought he was a genius."[92] Historians commonly see Rodgers as one of the important figures in the introduction of professionalism to the U.S. Navy. He was central in the establishment of common disciplinary procedures and, as part of the Board of Commissioners, helped develop early naval policy. He had his baptism of fire in the Quasi-War, serving under Capt. Thomas Truxton on board USS *Constellation*. While other junior officers like Lt. Isaac Hull and Midn. David Porter were conducting expeditionary raids, cutting-out expeditions, and convoy escort against pirates and privateers, Rodgers served in the major frigate duels of the war, *Constellation* versus *L'Insurgente* and *La Vengeance*, and as a part of squadron operations in search of French warships. In the years leading up to the War of 1812, he focused on developing standardized night signaling for squadron operations and the development of standardized rigging and mast sizes for the same classes of ships. He was a sailor with a focus on the blue water and conventional naval engagements. When it came time to oversee Robert Fulton's 1810 experiments in New York, he was predisposed to dislike the very idea of a torpedo, even without Fulton's aggressive claims that he would prove navies irrelevant.[93] His active campaigning against torpedo use, to include turning a blind eye to something that might have made a significant contribution at Baltimore in 1814, demonstrates the traditionalist archetype in the American officer corps.[94]

Stephen Decatur exhibited the opposite disposition. His father had expressed concerns over his son's "wild and unsettled character" early on in his career.[95] Like Rodgers, Decatur Junior began his service at the navy's establishment for the start of the Quasi-War. Serving on board the frigate *United States* under Capt. John Barry, however, Decatur saw no traditional naval combat. *United States* spent the war commerce-raiding and chasing privateers and never fell in with a French warship. Decatur's real combat experience did not come until the Barbary War, which, as illustrated in chapter 2, was made up of duty in gunboats and on raiding operations as much as blue-water service. When the British blockaded Decatur into New London in 1813, he became a confidant of Robert Fulton and a full supporter of the partisan operations that developed in Long Island Sound. He embraced naval irregular warfare and encouraged it, working directly with Fulton on the design of new weapons and platforms.[96]

Also in the vicinity of New York, Master Commandant Jacob Lewis was a regular supporter of irregular operations. He worked directly with the partisans who developed the schooner *Eagle*'s attack on *Ramillies*. Like Decatur, he worked with Fulton on designing new weapons. He also developed a number of his own tactical innovations, including joint operations pairing gunboats with militia units to attack British supply parties, and disguising vessels for countersmuggling operations. However, Lewis did not have a traditional naval background. He began the War of 1812 as the captain of a privateer, and after several very successful cruises, he applied to the secretary of the navy for a commission. Like his fellow privateer Joshua Barney in the Chesapeake, the secretary gave him a gunboat flotilla, and he proved to be an innovative officer. Without the support of officers like Lewis and Decatur, and Charles Gordon in the Chesapeake, the efforts of partisans would have faced much greater difficulty.[97]

Of all the episodes studied in this book, this chapter is the example most tied to a technological advancement in naval weaponry. Robert Fulton's development of the torpedo and undersea warfare are central to this history. However, operationally the technical aspects of the design and construction of the torpedoes were less relevant than the platforms from which the Americans launched them. The other key aspect to the new weapons' apparent failure is the link between people and platforms. Specifically, these examples show that training and practice are required at the tactical level for successful irregular warfare and to ensure that new technology is employed properly.

On Long Island Sound, the naval irregular warfare operations were ancillary to the commanders' larger intent. Decatur was constantly searching for a way to escape the blockade. His support of the partisan attacks on the British blockading squadron stemmed from his desire to change the operational dynamics in Long Island Sound and create an opportunity for his escape. Charles Gordon's experience with Elijah Mix on the Chesapeake, however, offers a counter example. Gordon saw Mix's attempts to develop torpedo boats as an opportunity, which he could use to enhance the operational capabilities of his small defensive squadron in the bay, integrating them directly. The attempt to convince the Baltimore Committee of Defense to help procure an old brig, as a ship to carry a number of torpedo boats, is one of the first examples of creating a mother ship to transport naval irregular warfare units. If Gordon and Mix had been able to convince the committee, the brig would have sailed with the schooner squadron, adding the ability to conduct irregular warfare strikes in addition to scouting and engagement of smaller British warships.

The torpedo operations conducted in the War of 1812 also illustrate that nexus of people and platforms involved in the training for and practice of naval irregular warfare. These missions generally involved operators who had neither trained for nor practiced the tactics necessary for successful torpedo operations. While the navy gave these groups logistical support, and sometimes operational advice, there was no attempt to practice the tactics involved either at the behest of the navy or in partisan groups themselves. The result is a number of near-misses and failed operations that may demonstrate the inexperience of their crews more than the failure of the theories behind the tactics and weapons. Fulton provided his pamphlet, which served as a tactical manual, but reading about how to conduct an operation does not automatically prepare one for the ability to execute it in real time. The inability to judge the speed with which the torpedoes would float with the tide, ignorance of proper settings for the timing systems used in the detonators, and lack of practice with the plunging boats all affected the success of the torpedo attacks.

For example, following Mix's attack on *Plantagenet*, his interview with the *Norfolk Herald* indicated he knew exactly why the torpedoes detonated early, and he would be able to correct the mistake. It is likely he misjudged the movement of the tide and set the timer in the torpedo detonator for too short a period. This mistake mirrored Fulton's experience in Britain during his demonstration against the brig *Dorothea*. In that exercise, he had his two crews assigned well before the event and he organized several days of practice with them. During this time, Fulton was able to get not only the hardware set properly, but the crew practiced in judging the current and tides off Deal and in placing the torpedoes properly for effective targeting. The result was a successful sinking. Naval irregular warfare tactics, techniques, and procedures are no different from other elements of warfare and require training and practice in order to ensure success.[98]

As with previous examples of irregular operations, the partnerships between distinct groups are a significant part of what made the plots and expeditions possible. Specifically, the work of Robert Fulton, both as a proponent of the technology and as a center of tactical knowledge, was critical. Officers and sailors of the U.S. Navy, commonly cast by historians as merely interested observers in the attacks, frequently played central roles in the logistics and organization needed for the operations. Fulton and the navy worked alongside the partisans themselves. Sometimes, the navy formally took responsibility for the missions. In other examples, as the Long Island Sound operations demonstrate, it only provided logistical support and advice. The combination of Fulton, the partisans, and the navy made partnership an important element of these irregular missions.[99]

The attacks led by Elijah Mix on the Chesapeake are an excellent example. Mix sent his proposals to the president and to the secretary of the navy, but he also contacted Fulton via mutual friend Benjamin Latrobe. While Secretary Jones ordered Gordon in Baltimore to provide Mix with five hundred pounds of powder, a fast boat, and six volunteers (if he could find volunteers), he also formally appointed him as a sailing master, in essence making his expedition a sanctioned U.S. Navy operation. Meanwhile, Mix signed a contract with Fulton in which the two of them agreed to split the proceeds from a successful attack if Fulton provided the latest torpedo technology and technical advice. The cooperation with the navy and Fulton contributed directly to Mix's operations. He also trained James McGowan in torpedo operations, expertise the midshipman would take with him to the Great Lakes the following year.[100]

Contracts between Fulton and the partisan groups were common. James Welden, who attempted the "plunging boat" attack in August 1813, was under contract with Fulton just like Mix, as were some of the partisans in the Delaware Bay expeditions. Fulton provided detailed instructions on the tactics, techniques, and procedures for the attacks, as well as the weapons themselves. His advice included the kinds of reconnaissance the attackers needed to conduct, how to rig their boats, and even how to dress the attackers to camouflage them at night. In return, he would share in the Treasury payment from any successful attack.[101]

If Fulton was the source of the tactics and the weapons themselves, the navy was commonly the logistical link in the chain. They provided boats in the Chesapeake and Delaware, as well as volunteers from the official rolls to fill out some of the crews. In Long Island Sound, Stephen Decatur and Jacob Lewis provided supplies, bases for operations, and advice for many of the operations, and were aware of others. On Lake Ontario, the expedition led by McGowan was entirely a navy affair, with uniformed sailors and government-owned boats, but built on the young officer's experience with Mix in the Chesapeake.[102]

During the War of 1812, torpedo-boat operations were in their infancy and their operational effectiveness was certainly marginal. The attacks did not sink any British warships, and the Treasury never paid out the rewards promised in the Torpedo Act. Yet how these attacks came together, what people were involved, which platforms were used to launch them, and the partnerships that contributed have had little study in the larger field of the War of 1812. These attacks certainly caused damage and loss of life, and they had an impact on the British methods of force protection and how they managed the blockade. Senior Royal Navy officers like Hardy, Cockburn, and Warren all worried about this new mode of naval warfare. It is easy to dismiss these examples because they never achieved their ultimate goal, the sinking of a British man-of-war. However, close operational-level study of the events shows they were in line with the larger principles of naval irregular warfare that had developed in the American navy in the Age of Sail. The torpedo operations, like the small-boat raiding operations on the Great Lakes, show that the operational history of the War of 1812 included more than dueling frigates and squadrons on inland lakes. After the war, the centrality of irregular operations would continue as the navy returned to the Caribbean to face the challenges of maritime security in a confused sea.

CHAPTER 6

PIRATES AND PRIVATEERS

Dawn of the West Indies Squadron

If the pattern of American history is a guide then after the Treaty of Ghent, which ended the War of 1812, the U.S. Navy should have retrenched and been significantly cut in size and scope. However, this was not the case. The War of 1812 appeared to end the long-running debate between the interest groups, which Craig Symonds categorized as the navalists and the antinavalists, in favor of maintaining a peacetime navy with capital ships and a global capability. Part of the reason for this was that, despite the end of the war, American foreign policy still appeared to need operational naval forces. During the war with Great Britain, the dey of Algiers resumed attacks on American shipping in the Mediterranean. The number of attacks by Algerian corsairs was limited, since the British blockade had curtailed American trade during the war. However, it was enough to gain the attention of the president and Congress. No sooner had the Senate taken up debate over ratification of the Treaty of Ghent but the navy was again preparing to deploy to the Mediterranean.[1]

The navy formed two squadrons under the command of Capt. Stephen Decatur and Capt. William Bainbridge. Secretary of the Navy Benjamin Crowninshield assigned the frigates *Guerriere* (44-gun), *Constellation* (36-gun), *Macedonian* (38-gun), the sloops *Epervier* (18-gun) and *Ontario* (16-gun), the brigs *Firefly* (14-gun), *Spark* (14-gun), and *Flambeau* (14-gun), and the schooners *Torch* (12-gun) and *Spitfire* (12-gun) to Decatur's force. They sailed on 20 May 1815, two weeks before Bainbridge and his ships. The squadron captured the Algerian flagship *Meshuda* and the brig *Estedio*, reaching Algiers with their prizes in company during the last week of June. Decatur began negotiations, with his guns run out and the city under the threat of bombardment from the largest squadron the United States had ever sent overseas. The dey, after venting his frustration to the British consul who had assured him the Americans would never respond to their attacks, capitulated within days and signed a new treaty with the United States.[2]

Bainbridge's squadron arrived too late to have any effect on Algiers, but his ships remained in the region as the navy re-established the Mediterranean Station, which it had formed as the first American overseas force following the Quasi-War with France. Decatur returned home with the majority of his ships. Bainbridge and the American squadron, led by the new ship of the line *Independence*, sailed southern Europe proudly demonstrating the naval prestige they earned during the war with Great Britain. American commerce expanded around the world as the restrictions of prewar embargoes and wartime blockades fell away and unleashed the American carrying trade. The U.S. Navy followed their merchantmen, with deployments to the Pacific, Indian Ocean, and the South Atlantic. After the experience of the war with Great Britain, and the challenges Secretaries Hamilton and Jones experienced in administering the service, Congress began taking naval policy and organization more seriously. In 1815, it established the Board of Naval Commissioners to advise the secretary of the navy and help administer the department. With the advice of the new board and the secretary, Congress then passed an "Act for the General Increase of the Navy" in 1816, which appropriated $10 million to build more heavy frigates and ships of the line.[3]

Despite building plans that aspired to a small squadron of ships of the line, which were on par with warships of the major naval powers of the world and would bring naval prestige to the United States, the next maritime challenge the United States faced was again an irregular one. Maritime insecurity, in the forms of smuggling, piracy, and the international slave trade, had been a long-running problem in the Caribbean and central Atlantic. Merchant traffic under the U.S. flag suffered from attacks in and around the Spanish colonies as early as 1791, when the schooner *Polly* reported being attacked on the north shore of Cuba.[4] But the long history of Caribbean piracy dated back into the seventeenth century, including what historians have labelled the Golden Age of Piracy from roughly the middle of that century to the middle of the eighteenth.[5] The U.S. Navy's operations against French interests in the region, outlined in chapter 2, were just the beginning of the U.S. involvement in Caribbean maritime security. In that conflict, the naval forces had to deal with not only commissioned warships and authorized privateers but also with pirates and smugglers who were breaking the U.S. neutrality statutes. The end of that conflict, however, did not end the challenges to the security of American shipping. In 1805, newspapers were again reporting depredations. The navy, in the post–Barbary War retrenchment of the gunboat naval policy, deployed the single corvette *John Adams*, under the command of Alexander Murray, but with little effect, despite President Jefferson's declaration that "our coasts have been infested and our harbors watched by privately armed vessels" preying on American merchant vessels.[6]

Even with the simmering of maritime insecurity to the south, the impact on U.S. foreign and military policy at large was limited. Following the Embargo Act of 1807, the Jefferson administration's own policies had reduced American overseas commerce. With Americans barred from overseas trade by their own government, not only in Europe but also in the West Indies and colonies, there were few reasons for American ships to set out to sea. During this period, those Americans who were attentive to foreign policy tended to focus on the looming issues with Great Britain that would finally come to a head with the War of 1812. However, other forces were at work within the Western Hemisphere that would lead to further maritime dangers.

In 1808, after two decades of conflict and struggle between the forces of revolutionary France and the established order in Europe, Napoleon sent an army into Spain. At first, they had permission to transit Spanish territory on their way to the occupation of Portugal. However, once the French army was on the Iberian Peninsula, Napoleon used its presence to claim control over both Portugal and Spain, which he had long desired to bring more directly into his empire. He lured the Spanish royal family into southern France for a diplomatic meeting and then forced King Carlos IV and his son, later Fernando VII, to abdicate. Napoleon replaced the Bourbon monarchy by making his brother Joseph the king of Spain. However, the rest of the Spanish Empire was not particularly happy with that result.[7]

In the Americas, colonial political leaders largely refused to recognize the legitimacy of the Napoleonic regime. The Spanish American colonies had experienced a measure of instability ever since their northern neighbors had rebelled in the American Revolution. In some cases, the instability was encouraged by British or newly independent American interests. However, following the abdication of the king to Napoleon, as historian Jaime Rodriguez O. wrote, "the people of the Peninsula and the New World . . . were virtually unanimous in their opposition to the French."[8] On the peninsula, local groups of Spanish political leaders began forming juntas to govern their localities in the name of the Bourbon monarchy until they could restore the rightful king. The New World followed the model from Spain. However, in the American colonies, local politics introduced divisions over exactly how the local leaders should run the juntas, and to what end. European Spaniards living in the colonies tended to want to find an entity on the Continent to pledge their allegiance to and from which they could take directions. The upper-middle-class Spaniards born in the Americas, known as *caudillos*, wanted to retain their own autonomy until Spain formally restored Fernando to his throne. These divisions sowed the seeds for a decade of conflict, the eventual push for full independence from Spain, and eventually the start of outright warfare.[9]

In 1810, the army that Napoleon had sent into Spain marched across the country with a string of victories against Spanish resistance. The Junta Central, which established itself in Seville to administer the empire, fell at the end of

January and Spanish loyalists retreated to the seaport of Cadiz. Word spread to the New World about the unravelling of the attempt to resist Napoleon on the Peninsula, which led a number of local Spanish American juntas to declare their autonomy. The first to do so was in the Viceroyalty of New Granada (which became known as Venezuela or Cartagena), followed by Rio de la Plata (also known by its primary city Buenos Aires).[10] This in turn led a number of provinces, including Upper and Lower Peru and Paraguay, to declare their autonomy from the Spanish American viceroyalties.[11] The results were the opening of what would develop into a civil war across Spanish America.

In 1813 Venezuela became the first of the revolutionary governments to issue commissions to privateers. Following in the wake of their northern neighbor, and the United States' successes with privateers in both the American Revolution and the ongoing War of 1812, the Spanish Americans looked for an inexpensive way to combat the regime in Europe and the Spanish colonies such as Cuba and Puerto Rico, who had not broken away. Revolutionary leaders granted these privateer commissions under the political authority of Cartagena, Venezuela, or Colombia over the course of the succeeding decades, depending on the state of the civil war and the political situation in what the Spanish called New Granada. Rio de la Plata followed suit in 1815, issuing commissions from Buenos Aires. The Mexicans and even the landlocked Paraguayans began assigning letters of marque and privateering commissions. The Spanish-speaking world called these privateers the *corso insurgente*, or the insurgent corsairs. In 1817, Cuba followed suit under the authority of the Spanish government and Puerto Rico soon began issuing commissions as well to combat the breakaway governments. Dozens of predatory ships were unleashed on the Caribbean world.[12]

For the first decade of the conflicts in Spanish America, the United States focused on its land borders and on the War of 1812. Specifically, with respect to Spanish interests, the concern was over East and West Florida, which the United States annexed over the course of that period through a sequence of diplomatic, military, and covert maneuvers (including the convenient use of filibustering).[13] The Adams-Onis Treaty of 1819, between Spain and the United States, formalized the addition of the Floridas to U.S. territory. During this period, naval leaders recognized the threat of maritime insecurity in the Gulf of Mexico and along the southern coast of the United States, but there was little acknowledgment of it from the wider public. The commanders of the New Orleans Naval Station during the first two decades of the nineteenth century, namely David Porter and Daniel Todd Patterson, struggled to impose order in American waters with a poorly manned and equipped force of gunboats.[14] When the Adams-Onis Treaty settled the land issue, Americans began to realize they had a secondary, maritime, crisis developing on their newly acquired southern shores.[15]

From 1817 to 1819, the number of attacks on merchant shipping by the Spanish American privateers climbed sharply. Peter Earle has suggested that the piracy and privateering in the early-nineteenth-century Caribbean was merely the extension of two centuries of maritime insecurity in the West Indies. But he admitted that "there was probably more piracy and maritime mayhem in the first fifteen years of what has been labelled the Pax Britannica than there had ever been in the so-called Golden Age of Piracy."[16] In his quantitative study of Spanish American prize actions, Matthew McCarthy demonstrated that the decades of the Spanish American wars for independence resulted in a dramatic rise in prize actions. McCarthy documented almost 450 attacks on ships of all nations during those three years, a part of 1,688 attacks in the West Indies during the decades of open warfare in the Spanish American colonies.[17] Newspapers in the United States regularly carried stories of violent attacks on American ships and international vessels, the more gruesome the details the better. With rising awareness of the threat, rapidly rising insurance rates, and regular calls on the government from merchant interests, Congress finally acted in March 1819 by passing "An Act to Protect the Commerce of the United States, and Punish the Crime of Piracy." The act authorized the president to deploy a suitable naval force into the West Indies and central and southern Atlantic, and to instruct naval commanders to seize any vessels committing depredations and send them into American ports for trial. It also clarified that it was legal for American merchant vessels to arm themselves for protection. Finally, it prescribed the death penalty for the act of piracy.[18]

What the Act of 1819 overlooked in many ways was the definitional difference between pirate and privateer. In large measure, this was the result of longstanding American policy on belligerent rights at sea, and the resulting legal conflict with much of the rest of the world. The United States believed in the protection of private property at sea and neutral rights. This was based around the idea that a ship sailing under the flag of a neutral nation was inviolable, and that the cargo in it was to be considered neutral by the flag of the vessel. Much of the rest of the world, led by Great Britain and including Spain and France, maintained the opposite: it was the ownership of the cargo and the cargo's destination that determined whether naval vessels considered a ship neutral. If the cargo was contraband of war, and was destined for an enemy port, these powers maintained they had the legal right to seize that property and the ship that carried it. The American slogan "free ships make free goods" was at the heart of the maritime legal, diplomatic, and military conflicts that the nation had been having with European powers for over two decades.[19]

During the American Revolution and War of 1812, American political leaders had authorized privateers to attack only vessels sailing under the flag of their enemy, Great Britain. While the United States Congress declared its "sympathy" for the Spanish American colonies in 1810, the formal position of the U.S. government, declared by President Madison in 1811, was neutrality in the conflict between Spain and its colonies.[20] Because of the American belief in neutral rights, this led to the conclusion that Spanish American privateers had no right to attack ships sailing under the U.S. flag, since they were neutral. Regardless of where the cargo was bound, if they had no right to attack a merchant vessel under a neutral flag any ship that committed such an "outrage" should be considered a pirate instead of a privateer, even if it had a privateering commission. The British, Spanish, and the insurgent governments took the opposite view, believing they had the right to take any ship they determined was carrying the enemy's cargo or contraband. In many ways, this offered Congress the opportunity to sidestep the fact that American investors had built and outfitted many of the ships sailing under insurgent commissions in the United States, something the Spanish government certainly did not ignore.[21]

In response to the congressional action in the spring of 1819, Secretary of the Navy Smith Thompson ordered Capt. Oliver Hazard Perry, the hero of the Battle of Lake Erie, to take the corvette *John Adams* and the schooner *Nonsuch* and sail for Venezuela. Perry's instructions were as much diplomatic as military, and for a portion of them Secretary Crowninshield placed him formally under the direction of the State Department.[22] His primary goal was to open a dialogue with political leaders in the revolutionary territories and negotiate an understanding between them and the United States over who was and who was not a privateer, and to insist ships sailing under their commissions adhere to strict regulations with strict punishments for violation. *John Adams* and *Nonsuch* arrived off the mouth of the Orinoco River on 11 July 1819, and Perry transferred into the shallower draft schooner to head up river to meet with the vice president of Venezuela at Angostura (modern-day Ciudad Bolívar). His negotiations were a relative success. They resulted in a promise that the Venezuelans would furnish an official list of the commissioned privateers to the U.S. government, an acknowledgement from the revolutionary government that they were obligated to regulate their vessels better, and the assurance that the Venezuelans would provide restitution for wrongful captures. Despite these successes, Perry's mission ended in tragedy. As *Nonsuch* departed from the revolutionary capital, the captain contracted yellow fever and his condition deteriorated rapidly. By the time the schooner reached the Caribbean, where *John Adams* lay anchored with a surgeon on board, it was too late for Perry and he died at sea on 23 August.[23]

The following month, after word of Perry's death reached Washington, the secretary dispatched Capt. Charles Morris, on board the frigate *Constellation*, for the South Atlantic to take up the remainder of Perry's mission with the *John Adams* in company. The corvette had sailed for home with the news of Perry's death, but the schooner *Nonsuch* continued south along the coast of South America. Morris and his ships rendezvoused with her at Montevideo, where he transferred on board just as Perry had in order to navigate upriver to Buenos Aires. What he discovered, however, was far different from what Perry found in Venezuela. The dissolution of the former viceroyalty of New Granada had been relatively orderly, with the revolutionary government assuming control. But in Rio de la Plata, things were deteriorating. While the government in Buenos Aires claimed control over the entire region, it was simultaneously fighting its own civil war with the provinces that wanted to break away. Morris discovered there was nobody he could negotiate with who had enough power to enforce an agreement. Empty-handed, Morris sailed for home to make his report to Secretary Thompson. The navy concluded the privateer commissions from the southern half of the continent did not have the same legitimacy as those granted in the north.[24]

By the end of 1819, besides the promises from Venezuelan diplomats to regulate their privateers more effectively, very little had been accomplished. In December, the leaders of six maritime insurance companies sent a letter to President James Monroe. They outlined "the late unparalleled increase of piracies," listing forty-four attacks within the past year, and stated their desire to help the government in any way they could, specifically abstaining "from offering any comments or opinions" on what the government had already tried. Throughout the letter's passive-aggressive tone, it was clear the men regarded the government's policy as inadequate.[25] In his study of the American naval response in the West Indies a century later, Caspar Goodrich observed that the naval forces out of New Orleans continued to make some captures and to patrol as best they could but "such small successes were without material influence on the situation wherein piracy was indistinguishable from privateering. The feeble efforts of our government were wholly inadequate to cope."[26]

Congress renewed the Act to Protect Commerce in 1820, extending it for two years.[27] But, despite the growing outcry and continued newspaper reports and opinion articles protesting of attacks in the Caribbean and Gulf of Mexico, the force deployed by the navy remained small and based out of New Orleans. There were some limited successes. The brig *Enterprise*, under the command of Lt. Lawrence Kearny, captured four pirate schooners in October and took another in December. The sloop *Hornet*, Capt. Robert Henley commanding, captured the pirate schooner *Moscow* in late October as well.[28] Despite modest naval successes, in 1821 and 1822 the number of attacks again attracted the attention of the American people and the government. This time, the reason was not the corso insurgente but instead the rise of Cuban-based coastal piracy as veterans of the Gulf of Mexico smuggling-piratical industry reinforced bands of local Cuban watermen.

The colonial government's control in the provinces outside Havana was limited, and much of the Cuban piracy appeared to be the spread of banditry onto the sea. But it was more central to the Cuban economy than the banditry label suggests. McCarthy has illustrated it as such: "Far from being outlaws living apart from established society as their eighteenth-century predecessors may have been, Cuban-based pirates were an integral part of the economic and political life in Cuba in the early 1820s."[29] The operational methods used by these pirate groups were similar to those seen across centuries: men armed with small arms and personal weapons filled small craft, and patrolled in the shallows and at choke points where merchant ships were vulnerable. Easily blending in with the maritime industry, appearing as fishermen when they hid their weapons, these bands took prizes and ran them into the shallow bays and creeks of Cuba, where they could plunder them. The pirates shipped the merchandise inland and resold it with the help of local merchants, the more valuable items finding their way to Havana and into the legitimate trading markets.[30]

In early 1822, with a fresh outcry in the press, the American government again acted. Congress authorized the formation of a West Indies Squadron and Secretary Thompson placed it under the command of Capt. James Biddle. Made up of the frigates *Macedonian* and *Congress*, the corvette *John Adams*, as well as the sloops *Hornet* and *Peacock*, brigs *Spark* and *Enterprise*, and schooners *Alligator*, *Grampus*, *Shark*, and *Porpoise*, the squadron mounted over two hundred guns and brought extra manpower to work the ships' cutters and boats.[31] Biddle first sailed for Havana to meet with Governor (and Captain General) don Nicolas Mahy and his Spanish naval commander. The captain's purpose was to establish an operational working relationship with the local authorities. He proposed an agreement to allow American sailors and marines to land on Spanish territory when in pursuit of pirates. The Spanish authorities, however, refused, stating they had already deployed sufficient forces to address the issue and cited their own sovereignty. Biddle, disappointed, returned to sea and began sending his ships on patrol and assigning them to escort convoys.[32]

The first year of the West Indies Squadron's existence met with a number of successes, but also revealed limitations in the U.S. Navy's approach to counterpiracy operations off Cuba and in the wider Caribbean. In early March, *Enterprise*, still under the command of Kearny, took eight pirate vessels ranging in size from launches to schooners, with several hundred captives. The *Grampus* and *Shark* each took a schooner in June, taking three prisoners as the rest of the pirates escaped overboard.[33] In August, the schooner *Grampus*, under the command of Lt. Francis Gregory (see chapter 4), spotted a large hermaphrodite brig while escorting two American ships in a convoy to Curacao.[34] When *Grampus* sailed for her, the vessel first raised British colors, then hauled them down, raised Spanish colors, and fired a cannon to windward, generally considered a sign of peace. The closer *Grampus* approached, the more Lieutenant Gregory believed the ship fit the description of one that had attacked the American schooner *Coquette* just a few days before. Once within hailing distance, Gregory called over to the vessel and demanded she allow a boarding party over. The ship answered with a broadside and full volley of musket fire, which *Grampus* returned, and a general action commenced that lasted just three minutes before the other ship "stuck his colors, a complete wreck." Gregory sent a boarding party and discovered the vessel was the *Palmayra*, also known as the *Pancheta*, a privateer commissioned out of Puerto Rico but with a reputation of running fast and loose with the rules of engagement set out in its commission. Gregory sent the ship to Charleston as a prize, since it had fired on a U.S. warship without provocation. Despite these successful operations by smaller ships, Commodore Biddle had no luck in the squadron's flagship *Macedonian*. The frigate cruised a route through the entire Caribbean, including the Greater and Lesser Antilles and the Spanish Main, and yet did not engage a single pirate or privateer. Many of the ships convoying merchantmen never encountered a threat. Working the open water and regular sailing routes seemed to have little impact, and the number of captures was small compared to the number of attacks recorded in the newspapers.[35]

The two most successful early operations, however, did not occur outside the reef, as most of the captures already listed did. Instead, in the fall of 1822 two American lieutenants began leading their men and boats inside the reefs and islands that line the north coast of Cuba, into the bays and estuaries the pirates used as bases of operations but which were also within Spanish territorial waters. The first was Lt. Stephen Cassin, who received intelligence of a pirate base in "Honda Bay" while he lay anchored at Havana.[36] He sailed his sloop *Peacock* fifty miles west to patrol off that part of the coast. The morning of 28 September, the lookouts spotted three small sails heading out from the bay toward the ship. As the Americans came about to give chase, the sails also turned and sought to run away. *Peacock* overtook and captured one of the ships, placing a prize crew on board. The Americans spoke with the small Royal Navy schooner *Speedwell*, which was also patrolling the region, and learned the British had been attacked that morning, but by a larger force of two schooners, and had been unable to get after their attackers once they returned inside the reef.[37] Cassin immediately manned his boats and sent them inshore to find the pirate camp he was sure was there. They returned at sunset without success. However, the next morning the small flotilla made up of *Peacock*, *Speedwell*, the captured prize vessel, and the American revenue cutter *Louisiana* spotted a sail inside the reef and again manned and dispatched their boats. They returned the following morning with four pirate schooners captured inside the bay.[38]

The second example was just over a month later. On 8 November 1822, *Alligator* came to anchor in Matanzas harbor, a popular trading harbor sixty miles east of Havana and almost due south from Key West. The shores around Matanzas and the small islands just off shore to the east had begun to earn a reputation as one of the common pirate havens. Lt. William Allen, the commander of *Alligator*, received word almost immediately upon anchoring that two American merchant vessels, a brig and a schooner, had been attacked about forty miles east at "Cape St Haycacos," leading into the Bay of Cardenas. The pirates released the captain of one of the ships and the mate of the other to travel to Matanzas in order to convey the pirates' ransom demands. They were given three days to return with $7,000 or the pirates would burn the ships and kill their crews. Allen brought both men on board *Alligator*, took a small schooner in company, which the local merchants had been fitting out to try and retake their ships, and departed the harbor.[39]

The next morning, *Alligator* and its consort were off the Stone Keys, about ten miles from where their merchant sailor guides remembered the attackers had hidden the vessels. With so much shoal water in the vicinity, *Alligator* came to anchor and the crew hoisted out her boats to head into the keys. Allen divided his crew among the boats. He and his marine commander, Capt. William Freeman, manned the schooner's launch with thirteen sailors and marines and brought along the merchant captain whom the pirates had sent for the ransom. The mate of the other captured vessel joined Lt. John Dale and a crew of ten in the cutter. A midshipman and four sailors manned the captain's gig. And finally, Allen placed the small schooner, which the Matanzas merchants had fitted out, under the shared command of Acting Lt. Robert Cunningham and its civilian master and manned it with a mixed crew of twenty sailors, marines, and merchantmen. Acting Sailing Master V. M. Randolph remained in charge of the anchored *Alligator*, with the remaining eighteen men as a skeleton crew.[40]

In the growing daylight, the Americans spotted several vessels lying at anchor just beyond the islands at the mouth of the bay: a ship, a brig, and three schooners. Further to leeward another small schooner, with its deck full of men, was under sail and small boats appeared to be moving between it and the anchored vessels. As two of the anchored ships prepared to get under way, the American flotilla set out. There was a slight wind, and the small schooner made little way as the sailors and marines rowed toward her. Spotting the Americans, the schooner ran out sweeps and her crew began attempting to pull away. However, *Alligator's* raiders continued to close. As the boats pulled within range, the schooner they were chasing rounded up into the wind as she hoisted a blood-red pirate ensign. Her guns came to bear on the boats and the schooner let loose a broadside of round and grapeshot.[41]

The Americans pivoted their boats, still out of range for their muskets and small arms to return fire, and maneuvered to keep themselves in the schooner's wake and thus away from the heavy fire of her main guns. As Cunningham attempted to bring the merchants' armed schooner around the point, the vessel grounded on a bar and was unable to get into the engagement. Allen led the attack with just the boats. About ten minutes after the first shots were fired, he had his launch and the cutter close aboard and began to take position to board over the schooner's quarter. A second pirate schooner, which had approached from the anchorage, crossed her bow and began firing. After a few last shots at the Americans, the pirates abandoned the first schooner by climbing into boats, which they had lashed on the schooner's starboard side. As they took their positions and started to cast off for their compatriots, Allen ordered his launch and the cutter in pursuit. With the Americans approaching, the pirates realized they would not be able to outrun the larger and better manned launch and cutter. They turned around and reboarded their abandoned vessel. The Americans maneuvered back to the stern, under fire from the second schooner, and commenced raking the deck of the first schooner with their musket fire. The pirates again elected to abandon ship, pulling hard at their oars to join their consort.

Allen ordered Midshipman Henley and his sailors in the gig, which trailed behind, to take possession of the prize schooner. With Lieutenant Dale and the cutter in company, he set off for the second schooner, now manned with between sixty and eighty pirates. They poured a withering fire into the Americans, who returned fire and rowed hard for ten minutes, gaining on the poorly rowed schooner. Casualties began to mount in the launch and the cutter. The cutter was the first to fall away, dead and wounded sailors and marines lying across its shot-away oars. Lieutenant Allen stood up to rally his men. Two musket shots struck him as he encouraged his men on and he collapsed into the bottom of the launch. Unable to reach the second schooner, the two boats withdrew to the safety of the prize vessel where Henley and his men had taken control. The sailors lashed the boats alongside and pulled the killed and wounded up on deck, including Allen who, despite reportedly keeping up a positive attitude for his men and continuing to provide orders, would soon succumb to his wounds.[42]

A third and final pirate schooner set sail from the anchorage, after cutting loose some of the captured merchant vessels. The pirate schooners sailed away inside the barrier islands. On board their prize, the Americans took stock of what they had accomplished. Dale, who had been Allen's first lieutenant, took command. The Americans discovered fourteen dead pirates, but there were likely others whom they had wounded in the fighting with the schooners but who had escaped, and Dale attributed some pirate casualties to a few who jumped overboard in an attempt to escape. In return, he had two dead of his own, three severely injured, as well as Lieutenant Allen and Seaman Elijah Place, who lay bleeding on the deck and would soon die from their injuries. The captured schooner was about 80 tons, mounted a single long 12-pound gun, two long 6-pounders, two 3-pounders, and two swivels, and had been manned by about thirty-five pirates. The two that escaped appeared to be about 60 and 90 tons each. On board the prize, they discovered the red flag was nailed to the mast, which was commonly a sign of pirates who would fight to the death despite the fact they had abandoned ship. It was possible one of the dead had been the commander who ordered the flag nailed so it could not be struck. Searching the cabin, the Americans also discovered several bottles filled with powder and corked with slow burning fuses. The fuses appeared to have been lit, but then put out again, and one of the crew surmised it must have happened when the pirates reboarded the ship before abandoning it the second time. Either way, the crew of the gig appeared lucky that the trap had not detonated.[43]

Dale sailed the prize schooner back to *Alligator* and anchored alongside. He then dispatched small crews, along with the merchant sailors, to begin collecting the vessels left behind by the fleeing pirates. The crews brought out the ship-rigged *William Henry* of New York, the brigs *Sarah Morril* and *Iris* of Boston, and a small schooner out of Rochester and another from Salem. With the merchant schooner under Cunningham's command recovered off the bar on the rising tide, the American ships returned to Matanzas. William Allen was buried in a ceremony attended by a number of local merchants. In Matanzas, the Americans also learned that Spanish authorities had captured a number of pirates who attempted to escape ashore during the engagement. Dale sent the recaptured ships and the pirate schooner, named *Revenge*, back to the United States under the command of marine Captain Freeman for adjudication in court.[44]

The memory of Lt. William Allen's aggressive leadership became an inspiration for the junior officers serving on the West Indies station.[45] As the news of the engagement and Allen's death spread, President Monroe continued corresponding with Congress on the threat to American interests. On 10 December, he wrote to the Senate: "Recent information of the multiplied outrages and depredations which have been committed on our seamen and commerce by the pirates in the

West Indies and Gulf of Mexico, exemplified by the death of a very meritorious officer, seems to call for some prompt and decisive measure on the part of the government."[46] Following the president's letter, Secretary Thompson took up the issue with the Board of Naval Commissioners, asking them for advice on how to best combat the threat. Captains John Rodgers, Isaac Chauncey, and David Porter examined the examples offered by both Lieutenant Allen and Lieutenant Cassin and reassessed the strategy the navy was pursuing off the coast of Cuba.[47]

The operational record of the first season the West Indies Squadron spent in the Caribbean demonstrated that the larger ships had far less success than the smaller ships. The 12-gun vessels had the most success, with *Enterprise*'s captures in March, *Grampus*'s capture of *Panchetta*, and *Alligator*'s capture of *Revenge* and recapture of five prizes. *Porpoise*, an 18-gun sloop, was on the larger side of the ships in the squadron but still considered small by the navies of the day. Even then, Cassin's operation used the British tender *Speedwell* and the smaller revenue cutter *Louisiana*. The heavier ships, the frigates *Macedonian*, *Congress*, and corvette *James Adams*, had almost no apparent success. The Board of Naval Commissioners developed a new plan for the West Indies Squadron. The makeup of the squadron from 1823 onward was informed by David Porter's experience in the Pacific in the War of 1812, where he used captured whalers to create a squadron of small raiding ships to attack British commerce off the South American coast and in the islands of the South Pacific. In December of 1822, Porter resigned his position as a commissioner and accepted President Monroe's offer to command the new West Indies Squadron.

The "Mosquito Fleet," as the officers and men nicknamed the new conception of the counterpiracy force, was biased toward small ships. Many of the most successful and fastest privateers, both under U.S. commissions in the War of 1812 and under the commissions of the Spanish American revolutionary governments, were from Baltimore. The heart of the new force was ten schooners, built on the Baltimore clipper style, with less than 7 feet draft, a single long 12- or 18-pound gun on a pivot, two 12-pound carronades, and a number of small arms and swivels. The schooners were fitted with sweeps to propel them in a calm. In addition, the commissioners suggested five large cutters, or barges, with twenty oars and enough room for forty men armed with muskets, pistols, cutlasses, and pikes. The final and perhaps most innovative part of the squadron was the introduction of a steam vessel. The plan called for "one steam-boat, of ninety to one hundred and twenty tons, to carry two eighteen pounders and two twelve pounders upon traveling carriages so as to fire from any part of the vessel."[48]

Working inside the reefs and islands of the Cuban coast required a lot of rowing and long periods in open boats that could exhaust the crews. A paddle-wheel steamer could tow the barges in close to their targets for planned raids, and propel the crews while patrolling in the unreliable winds so they could save their effort for pursuing suspected vessels. The navy's first experiment with steam had been the *Demologos*, designed and built as a floating battery by Robert Fulton during the War of 1812. Renamed *Fulton* after its designer's death, the navy commissioned the ship in 1816, but almost immediately placed it in ordinary and used it as a receiving ship for the rest of its life.[49] Porter found a side-wheel steamer named *Enterprise* built in Hartford, Connecticut, in 1818, which was serving as a ferry boat. The navy purchased the 100-ton craft, with the essential wood-fueled copper boiler, for $16,000 and rechristened it *Sea Gull*. It became "the first steam-propelled man of war engaged in actual warfare," since it would come under fire and see combat repeatedly in the coming years of irregular warfare.[50]

The navy estimated the Mosquito Fleet's ships would cost $44,000 and they projected the total cost, including manpower and supplies for a year, to be $115,308, which was less than the cost to build a single frigate, never mind the cost to man and operate it.[51] Congress still saw it as too much money and appropriated funds for the steam boat, five barges, and only eight of the schooners instead of ten. The new squadron sailed south on 14 February 1823. They established a base at Thompson's Island, known today as Key West, Florida, which the sailors also called Allenton in honor of *Alligator*'s commander, and gathered the sloops, brigs, and schooners that remained in the region.[52]

When he first arrived at Puerto Rico and Cuba, Porter discovered that he had a significant diplomatic challenge to overcome. His interaction with the leaders of the colonies suggested that both islands suspected he had disguised the American force under his command as a counterpiracy squadron and instead it was actually an invasion force.[53] It was not an unreasonable concern, considering how the United States had obtained Spanish Florida just a few years earlier. Porter attempted to allay the Spanish fears, especially after an incident in San Juan harbor when an excitable gun crew in the harbor fortress fired on *Fox*, killing Lt. William Cocke. Porter corresponded with both governors, assuring them the Americans did not intend to take Spanish territory.[54] While they seemed to take his assertions at face value, Porter also received the same reply as Biddle when he suggested cooperating and allowing Americans to land on Spanish soil. The governor of Cuba did issue a general order, instructing his forces to cooperate with the Americans in every way "which may be compatible with territorial privilege and respect."[55] For the officials of an empire, which faced threats from a multitude of internal and external interests, the response was the only one that made sense. Americans had encroached on Spanish sovereignty more than once in the previous three decades. Yet the rules of engagement authorized in Porter's orders from Secretary Thompson allowed his men to land if in hot pursuit of pirate crews, so long as Porter made the effort to warn the Spanish authorities that they intended to do so.[56]

The first year of the Mosquito Fleet operations proved the effectiveness of deploying a larger number of small, fast ships instead of frigates or corvettes. Where Biddle's ships had convoyed merchantmen and patrolled from port to port, Porter's squadron embarked on a specific campaign. In the opening months, the commodore divided his barges and half of his schooners into two divisions and sent one to patrol the north shores of Cuba, along with *Sea Gull*, and the other the southern coast. The remainder of the schooners continued to work with American merchants through convoys and frequent port visits to Matanzas, Havana, San Juan and the main commercial ports. Patrolling both outside the reefs and cays, as well as inside the shallows, the small-boat divisions searched for and discovered hidden ships, pirate camps, and stripped-down prizes. During long days in the sun, and with hard rowing, the American forces raided with repeated success and few casualties. Porter continued his diplomatic correspondence with the Spanish authorities and opened up a new channel of partnership: direct cooperation with the Royal Navy.[57]

Lieutenant Cassin's operation with *Speedwell* in September 1822 foreshadowed this cooperation. Anglo-American relations remained tense; only a few years before they had been at war. Each remained suspicious of the other's intentions in the Caribbean. When Porter's squadron sailed from the United States, the British shared many of the same fears as the Spanish authorities about American activity. Secret intelligence reports to Vice Admiral Charles Rowley in Jamaica, from both the British consulate in Norfolk, which watched the American naval base, as well as from Washington, suggested he needed to be wary of the squadron.[58] However, when Porter and Rowley began communicating they appear to have managed to allay each other's fears.[59] The internal and external correspondence between the Americans and British indicate trust grew between them after the sloop of war HMS *Bustard* arrived at Key West in late July, with a large outbreak of yellow fever. Porter immediately offered aid, and treated Captain Rawdon Maclean and his crew with "various acts of kindness." The American surgeons nursed the sick British sailors back to health and Porter offered medical supplies, fresh produce, sheep, and fish.[60] As the Anglo-American relationship strengthened, Porter and Rowley coordinated how they would focus their patrols, based on the location of their bases at Key West and Jamaica. Porter concentrated his squadron's operations on the northern coast of Cuba and Puerto Rico while the British, with their base in Jamaica south of Cuba, sent their patrols primarily to the southern coast.[61]

The biggest challenge the American squadron faced was nonoperational. While the junior officers, sailors and marines sent inshore met with repeated success, the environment and disease became the biggest threats. By August, after several months of physically demanding work in the Caribbean's summer heat, yellow fever spread through the squadron, reaching the base at Key West. Doctors and scientists did not discover the cause and proper treatment of the fever for decades, hampering American military operations repeatedly from the Civil War through the Spanish-American War.[62] Porter reported to the new secretary of the navy, Samuel Southard, "the yellow fever has lately made its appearance among us to an alarming degree, and has carried off several."[63] He included stark medical reports from his surgeons noting eleven dead and twenty-one officers and sailors afflicted, including Porter himself "to a debilitating degree." Secretary Southard dispatched the schooner *Shark*, which had been refitting at New York, with Capt. John Rodgers of the Board of Naval Commissioners and "three of the most skilful [sic] and intelligent surgeons" in the service for Key West.[64]

The 1823 campaign against privateers and pirates in the West Indies was, for all intents, over. Porter's medical officers sent him north on board *Sea Gull*, fearful for his life. Before Rodgers and his surgeons completed their transit south, Porter ordered his heavier vessels north to the eastern seaboard for refit, and to allow his sailors and marines to recover. The squadron collected the healthy men together on board some of the schooners and left them in the region to provide convoys and a presence, to try and allay the fears of the merchants for the remainder of the year, resupplying as needed out of Matanzas rather than the "pestilent" Thompson's Island.[65] Despite the rapid and dramatic end to the 1823 campaign, the West Indies Squadron, operating the mosquito fleet concept of operations, had experienced significant success. Secretary Southard wrote in his end of year report, "Piracy, as a system, has been repressed in the neighborhood of the island of Cuba and now requires only to be watched by a proper force."[66] President Monroe's message to Congress at the end of the year lauded Porter and his men: "[T]he piracies by which our commerce in the neighborhood of the Island of Cuba had been afflicted have been repressed, and the confidence of our merchants in great measure restored."[67] Diplomats in the region and the newspapers seconded the sentiments that winter, but many feared that when the force left the pirates would reappear.[68]

———————

Those fears were well founded. Attacks resumed in the late autumn and early winter as the naval forces left behind operated without any real command and control and had only a limited presence. Newspapers reported the attacks and word spread in Washington, where Porter was recuperating, that the threat had rebounded.[69] In mid-December, after the president's annual message made the rounds in the nation's newspapers proclaiming success, Porter sailed down the

Chesapeake Bay for Norfolk to oversee the repair and refitting of his ships before returning to the Caribbean. In January, he received new orders from Secretary Southard expanding his responsibilities to include more attention to the Mexican coast, which had previously been under the purview of Patterson at New Orleans. The new orders also included the west coast of Africa and counterslaver operations in his mandate.[70] The navy adjusted the makeup of his squadron as well. Porter took *John Adams* as his flagship instead of the sloop *Peacock*, the larger vessel being better suited for logistical purposes rather than an operational role. Porter intended to leave the corvette at the base at Thompson's Island for the majority of the year, reassigning the vessel's manpower to augment the schooners and man the barges and boats, while the stowage, staterooms, and berth decks would be the squadron's expeditionary base. The sloop of war *Hornet* and brig *Spark* were joined by eleven schooners: the eight Porter had purchased the previous year as well as *Grampus*, *Shark*, and *Porpoise*. *Sea Gull* also returned to the Caribbean, as well as the storeship *Decoy*.[71]

Porter departed Norfolk at the end of January 1824 on board *John Adams*, with *Grampus* and *Sea Gull* in company. He had already dispatched schooners to begin patrolling off Cuba and Puerto Rico and initiated a rotation of the small ships off the Mexican coast, as his new orders required.[72] In February and March, the squadron returned to the West Indies and began scouring their previous hunting grounds. Porter took *John Adams* to make calls at numerous islands in the Greater Antilles, including St. Barts, St. Kitts, St. Thomas, and the frequent pirate and privateer havens in the cays south of Puerto Rico and the Mona Passage. He also stopped at Kingston, Jamaica, and communicated with Commodore Edward Owen, who had replaced Admiral Rowley as the commander of the British squadron in the middle of Porter's 1823 operations, and then to Havana to ensure the Spanish authorities knew he was back in theater. From Owen he learned that piracy on the southern coast of Cuba had revived as well. The British had sent patrols and launched operations against the pirates, suffering some casualties in the process. During his patrol, Porter discovered evidence of pirates, including camps that appeared recently abandoned, but made no captures. From Havana, he convoyed merchantmen to safety and then headed for Thompson's Island to reassign his crews to the barges and small craft.[73]

In the spring and early summer of 1824, the American squadron continued patrolling inside the reefs of Cuba and convoying merchant vessels. While they made some captures, the second year of Porter's command did not see as many engagements as the first year. Intelligence indicated some of the more notorious bands may have departed for the Yucatan coast of Mexico.[74] But by late May, yellow fever appeared again at Thompson's Island. This time the acting surgeon's mate leading the medical department was struck down and became so infirm he had to be sent north. Porter decided his squadron was not going to be caught out by the fever again or suffer the casualties he had the previous fall. He dispatched a letter to inform Southard he intended to order the majority of his force north by the middle of June. He also corresponded with Vice Admiral Lawrence Halstead, who had taken command at Jamaica from Owen, to maintain coordination with the Royal Navy's efforts.[75] He followed through on his plans to return home, arriving in Washington on 25 June, before the administration had received his communication informing them of the decision. Porter's nine-day transit, made possible by the direct route taken by *Sea Gull* rather than tacking into the Atlantic and back as the sailing ships did, resulted in him beating his own letter to the capital.[76]

Porter attempted to command the remaining elements of his squadron from Washington. However, doing so raised the indignation of Secretary Southard, and eventually President Monroe. Southard began sending Porter copies of every letter he or the president received reporting piracy and attacks off the Mexican coast, which included those from merchants in Matanzas who complained they were no longer being protected, as well as correspondence from the owners of insurance companies. Southard also started giving explicit instructions of how Porter should deploy his ships.[77] Porter took the reports, and the manner in which Southard forwarded them, as an affront. In August, Porter countered what he saw as an attack on him and his men by pointing out the success of the previous year, the laurels they had received, and the hard work of his officers and sailors.[78] But the secretary and the commodore continued to snipe at each other in their correspondence over the next two months. Finally, on 14 October, Southard wrote to Porter, "[I]t is deemed expedient by the Executive [meaning the president] that you proceed as speedily as possible to your station, in the *John Adams*, that, by your presence there, the most efficient protection may be afforded our commerce."[79] A week later, with no evidence of Porter's departure, the secretary demanded, "The presence of a commander on the station being now indispensable, you will proceed to it."[80]

Porter sailed for the West Indies in *John Adams*, and into his undoing. He made landfall on 12 November at St. Thomas, where he discovered news of an episode at the Puerto Rican harbor town of Fajardo.[81] When *John Adams* arrived, *Beagle* lay at anchor and Lt. Charles Platt visited his commander to relate the events of the past few weeks. In response to accusations made by an American shop owner in St. Thomas that Spaniards had robbed his store and taken goods to Fajardo, Platt had anchored *Beagle* in the Puerto Rican harbor. He went ashore to inquire about the stolen goods, electing to wear civilian clothing instead of his uniform. After he had discussed the robbery with the captain of the port and the *alcalde* (municipal magistrate) and explained that he was in civilian clothes because it was a sensitive civil matter, the Spaniards ordered Platt arrested. Even when the lieutenant sent for his uniform and his commission, the authorities "pronounced the commission a forgery and me a damned pirate, and ordered me to be confined in the jail." *Beagle* then sent more documentation ashore, including the lieutenant's orders from Porter, and eventually the authorities released him. Unsure what to do, Platt ordered his ship to weigh anchor and they returned to St. Thomas.[82]

The morning after his arrival in theater and the briefing from Platt, Porter ordered *John Adams*, *Beagle*, and *Grampus* (which had joined them at anchor) for Fajardo. Indignant over what he considered the personal affront to Lieutenant Platt, and the national insult to the U.S. flag by imprisoning him, Porter's force anchored off the Puerto Rican town and armed and manned their boats. Porter landed with two hundred sailors and marines and marched his force into town. Threatened by one of the ship's armed barges, Spanish soldiers abandoned their battery overlooking the port and the force landed without incident. Porter demanded an apology from the alcalde, under the threat that he would commence "the total destruction of the town" if he was resisted. The Spanish authorities quickly acceded to Porter's demands, insisting it was an unfortunate misunderstanding, and made an apology, which satisfied the commodore. Porter marched his men back to their waiting boats and they re-embarked.[83]

The commodore thought little of the Fajardo operation, moving on to restoring his local control of the squadron and deploying his ships to tamp down the occasional flare-up of prize actions and pirate attacks. Yet Southard and the president reacted differently when they received his report of 15 November. Already frustrated with him, Southard was smarting after Porter's combative correspondence during the late summer and autumn and the commodore's dubious request for relief. The landing at Fajardo was not a raid in hot pursuit of pirates, but an invasion of the sovereign territory of a foreign power with the intent to coerce a government. It was the last straw for Southard, who immediately relieved Porter of command and ordered him home, instructing him to turn over his papers and responsibilities to Capt. Lewis Warrington, who would sail shortly.[84]

The aftermath of Porter's relief, and the subsequent court of inquiry and then court-martial, has been widely examined by historians. However, the arguments involved in that process are less important to this study of the counterpiracy campaign and naval irregular warfare than the fact that it happened. Porter launched an operation that his civilian superiors believed overstepped a boundary, through his overt threats directed at the Spanish authorities on Puerto Rico. Historians in the first half of the twentieth century perpetuated the idea that disciplinary action against Porter was based on his operational decisions in Puerto Rico.[85]

This was exactly what Porter wanted, if his defense in court is to be trusted. More recent scholarship, and a review of the records of the proceedings, demonstrates the court found far less fault with his command of the squadron and actions on the beach at Fajardo than it did with his insubordinate correspondence and actions toward the president and navy secretary during the summer and autumn of 1824 and early 1825.[86]

At the end of January, Porter began turning over his responsibilities to Warrington and offered the relevant correspondence and orders he had given, as well as a summary of the squadron's status.[87] Warrington inherited a command that already knew its business. Many of the lieutenants and sailors had been serving in the West Indies throughout the years of the counterpiracy operations and the cruising grounds were familiar to them. Since they were patrolling inside the reefs and islands in the barges or with *Sea Gull*, or sailing outside in the schooners and landing occasionally to inspect fishing villages, the pirates and local authorities felt the American presence from Puerto Rico and St. Thomas across the Caribbean to the Yucatan peninsula and the coast of Mexico. In early March, Warrington transferred his flag to *Sea Gull* to take part in patrols himself, telling Southard that "nothing material has occurred on this station. No piracies have been committed for several months." The biggest news in the first months of his command had been the grounding and loss of *Ferret* in February, not unexpected with much of the squadron conducting inshore operations.[88]

But Warrington's report was too optimistic. Just the next week Lt. John Sloat, commanding *Grampus*, received reports of attacks off Fajardo. The Danish governor of St. Thomas offered Sloat a pair of small shallow-draft sloops, which he manned from his crew. The three ships set off in pursuit. The borrowed sloops patrolled past the barrier islands of Puerto Rico and into the Boca del Infierno, a bay on the southern shore west of Guayama. A pirate sloop engaged one of them, *Dolphin* commanded by Lt. Garrett Pendergrast, and exchanged fire with the Americans for about forty-five minutes before running itself ashore. The pirates jumped overboard and escaped into the brush, but the Americans floated the prize and brought her out without suffering any casualties.[89] Sloat had the cooperation of the local Spanish authorities at Ponce, whose soldiers captured a number of the pirates who tried to escape ashore. In a softening of the relationship between the Americans and Spanish, the captain general of Puerto Rico sent *Grampus* a letter of thanks and appreciation for their cooperation, recognizing the crew "who fought with so much bravery against the pirates."[90]

While Sloat and *Grampus* worked the waters around St. Thomas and the southern coast of Puerto Rico, Lt. Isaac McKeever was commanding *Sea Gull* off Matanzas, which remained a busy port for American merchantmen.[91] On 19 March, Warrington ordered McKeever to leave Lt. Thomas Paine and the schooner *Terrier* off the harbor's approaches and begin a patrol to the east past Stone Key and through the bays and reefs around the Camaguey islands. These were the same waters where Allen and *Alligator* had captured *Revenge* and recaptured the pirate prizes two and a half years earlier. McKeever took the barge *Gallinipper* in tow and got up steam on board *Sea Gull*, heading east through the Florida straits the next morning.[92]

Late in the day, *Sea Gull*'s lookouts spotted a large warship on the horizon near Stone Key. As they approached, the British colors stood out in the breeze and McKeever came alongside the frigate HMS *Dartmouth* out of Jamaica. He traded intelligence with Captain James Ashley Maude and learned the frigate was the flagship of a small force on counterpiracy patrol. The schooners *Union* and *Lion*, and some of *Dartmouth*'s boats, were working to windward on a patrol. McKeever steamed on, falling in with the British expedition the next evening in Cadiz Bay. The Americans came to anchor alongside the Royal Navy's ships and McKeever met Lieutenant Henry Warde, *Dartmouth*'s first lieutenant, to compare notes. The British had no specific intelligence of the location of a pirate camp, just a general report that two pirate vessels had been active in the area, but Warrington and McKeever had information of where camps had been set in the past and locations of common rendezvous for local pirates. Since they were patrolling the same area, and the waters further east were shallow, McKeever suggested the two expeditions join into a combined operation and leave the larger vessels behind. Warde agreed and the next morning *Sea Gull*'s crew manned *Gallinipper* and put out two cutters, while the British manned their own barge and two cutters. The combined patrol, which was flying the U.S. and British flags under McKeever's overall command, "cheerfully acceded to" by the British according to the lieutenant, set off inside the reefs and keys toward Sagua la Grande.[93]

The expedition faced contrary winds and a storm, which caused the lieutenants to pull their boats ashore for forty-eight hours and wait for the weather to pass. On the morning of 25 March, after the winds began to abate, the barges and boats reached the mouth of the Sagua la Grande River. McKeever found a local fisherman whom he compelled to pilot the party to a small island known to them as Juita Gorda Key, which was one of the locations Warrington had told him to search. Around four o'clock that afternoon the boats spotted a mast hidden in the mangrove bushes and began maneuvering through the narrow channel to find the ship. They approached a small schooner, not much larger than the *Revenge* captured by Allen and his men, pulled into a hidden anchorage. As the combined expedition approached within hailing distance, the schooner ran up

Spanish colors and its crew, who appeared busy offloading cargo, ran to quarters. With matches lit and cannon trained on the approaching cutters, the captain shouted to the Anglo-American boats to keep away or he would fire into them. A sandbar without a clear channel through it blocked closer approach to the schooner. As the cutters maneuvered and repeatedly grounded, attempting to find their way, the Spanish captain ordered his men to fire. They laid matches to the swivels, but the guns misfired. When one of the British boats crossed over the bar, McKeever ordered his men over the sides of their grounded boats. Splashing into waist-deep water, the American sailors and marines landed and took up firing positions on the beach sixty feet from the ship to create an enfilading crossfire from where the British boat floated, the Royal Marines' muskets also trained on the schooner.[94]

The captain of the schooner continued to shout threats. McKeever ordered him to strike his colors and come ashore, and not to consider trying to fire his guns again "at his peril." After the guns had misfired, the Americans reported the deck of the schooner was a mass of confusion. The captain agreed to come ashore and he and one of his men climbed over the side. But as they reached the beach, the two men raced off toward the mangroves. The Americans pursued, apprehending the captain while the other man escaped. At the same time, the British cutters, which had crossed the sandbar under the command of Master's Mate George Layall, came alongside the starboard quarter of the schooner. The marines brought the captain to McKeever, who ordered him to tell his men to strike their colors. Surrounded by what must have appeared a fearsome, sunburned, and wild-looking group of raiders flying the American and British flags, the captain relented and shouted orders for his men to haul down the colors. As one of the Spaniards worked the halyard, and the flag came down, however, a musket shot rang out from the schooner in defiance.

The Britons and Americans returned fire from their positions afloat and ashore. In the confusion of the firefight, the captain again tried to make his escape, but McKeever shot and wounded him with his pistol and he was taken back into custody. As some men from the schooner attempted to jump overboard and escape into the bushes, they were shot down. A boarding party under Master's Mate Layall swarmed over the sides of the ship and across the deck, driving the remainder of the crew below where they surrendered.

After securing the prisoners and searching the ship, the party discovered explosives that the pirates had attempted to rig, this time "lighted cigars in and near the magazine," which they extinguished. They took nineteen prisoners and recovered eight bodies and five wounded, but both lieutenants reported there were other casualties, who had either floated away on a strong tide or were dragged into the bushes by those who escaped. From its papers the schooner was determined to be *El Socorro*, armed with two 6-pound guns on pivots, four swivels, small arms, and rigged to sail with about a thirty-five-man crew, though the expedition believed there were more than that on board. Her hold was full of cargo, which the Americans and British believed was plunder since it was so fresh, including many valuables, and the crew had been offloading it into a secret cache hidden deeper on the island. The American and British search identified the captain as Antonio Ripol, and the search turned up papers that authorized the vessel to cruise for a month in the name of the local municipal government for the protection of trade.

Captain Maude, in his report to Admiral Halstead from *Dartmouth*, explained that the British had seen a number of these alleged "commissions" but treated them skeptically. Higher authorities in Havana never signed them, neither the governor nor military officials, and he firmly believed they mostly served as cover for piracy and complicit local authorities. In the case of *El Soccoro*, the commission was from the commandant of the Cuban town of Trinidad, who did not appear to have such authority. In his dispatch to the Admiralty forwarding Maude's report, Halstead agreed these documents were mere cover for pirate bands who had corrupt officials in their pockets. Warrington wrote nearly the same observation to Secretary Southard a month later.[95]

McKeever and Warde left three of the boats and a little less than half their party with *El Socorro*, under the orders of the American lieutenant Robert Cunningham, and took the remaining three boats full of sailors and marines to patrol to windward in search of the other pirate vessel the British had heard about.[96] After only a few hours, the expedition spotted sails—a large launch with a schooner rig—across the bay, and set out in pursuit. The boats gained on their quarry, which tried to run before the crew decided they could not escape. Instead, they ran the vessel ashore on a mangrove island and abandoned ship, disappearing into the swamp. The Americans and British boarded and discovered that, once again, the crew had attempted to rig an explosion; this time leaving a keg of gunpowder open next to the schooner's cooking fire and more powder trailed around the ship. They tossed the powder overboard and put out the fire. As they searched the ship, one of the Americans recognized it. Prior to joining the navy, Daniel Collins had served as second mate on board the merchant brig *Betsey*, which pirates had attacked in late November of 1824. In that particularly brutal attack, Collins had been the only member of the crew who managed to escape. He reported to McKeever and Warde that the launch they had just captured was the same vessel that attacked his ship and killed his shipmates the previous fall.[97]

While the British offered to take custody of the captured pirates, on McKeever's instructions (which were in keeping with American policy), the prisoners were turned over to the Spanish authorities at Sagua la Grande. The American boats regained *Sea Gull*, and the combined force rendezvoused with *Dartmouth* near Stone Key on the evening of 28 March.[98] Each report that made its way up the chain of command lauded the other nation's forces, making it clear it was a very successful combined operation. Warrington wrote to the secretary, "[W]e are under no small obligations to Lieutenant Ward [*sic*], the officers and men of H.B.M. ship *Dartmouth*, for their efficient co-operation, and their strenuous endeavors."[99]

———————————

After the success of the combined operation, the West Indies squadron continued to patrol but their presence appeared to put a lid on much of the lawlessness. In April, Warrington wrote to the secretary that the success of American operations had forced the remaining pirates to change their methods. Instead of schooners or medium-sized vessels that could work offshore, they resorted almost entirely to small boats. Hidden ashore or in creeks, they waited until spotters sighted a merchant vessel close enough to rush out and attack, but also made sure the horizon was clear of any other sails that might be naval patrols. The raid that followed was as fast as possible, and once the band plundered the target they moved along to a different location so there was no base or camp to help authorities trace them. Warrington made the point that such attacks were not going to respond to increased patrols. It did not matter whether they were inside or outside the reefs, only law enforcement ashore where the pirates sold the goods, or where the pirates lived, would be effective. While the American squadron had achieved a great deal, and reports of attacks on Americans were at an all-time low, there was little more Warrington could do without authorization to send troops ashore and conduct law enforcement instead of the Spanish authorities.[100] It seemed neither he nor the leaders in Washington were willing to go that far. Greater cooperation from the Spanish in both Cuba and Puerto Rico had been a central element to the development of policy in 1824, under Warrington's command. Occasionally the corso insurgente from Colombia or elsewhere still made an appearance, but they were the kind of threat the West Indies Squadron could handle. By August, Warrington reported to Washington from Havana, "If pirates are now or have been in force lately on either side of Cuba, they have not only abstained from making captures, but they have concealed themselves so effectually as to prevent detection."[101]

Secretary Southard and Commodore Warrington appeared to have derived solid lessons from the five years of the operations on the West Indies Station. Constant presence was required, as Porter's returns to the United States and the flare-ups that followed demonstrated. On top of that, the continued appearance of yellow fever at Key West suggested it was a poor base for operations. New

Orleans, however, required a transit up the Mississippi River, which placed limits on larger vessels. Instead of those locations, after conducting a survey of the Gulf Coast Warrington and Southard settled on Pensacola as the new base for the West Indies Station. The U.S. Army had a small garrison there at Barrancas, which they turned over to the navy, and Capt. William Bainbridge and Capt. James Biddle helped Warrington get the naval base established. Pensacola would prove a vital base for the squadrons that operated on the station in the coming decades, its importance illustrated by the role it played at the start of the Civil War. The West Indies Squadron became a permanent fixture of the post-1812 Navy, joining the deployments of squadrons to the Mediterranean and Pacific Stations.[102]

In his first annual message to Congress as president, John Quincy Adams praised Warrington's squadron and their efforts. But Adams, ever the pragmatist, made sure Congress understood the importance of the permanent presence Warrington and Southard had worked to put in place. He wrote that the safety of American commerce "for years to come" would be reliant on "the steady continuance of an armed force devoted to its protection."[103] Congress agreed and continued to fund the West Indies Squadron. The following summer Warrington reported from Havana, where he was on board his new flagship *Constellation*: "[N]o piracies have been committed since my last letter. Depredations of our commerce are fortunately unheard of where they were formerly so frequent."[104]

The counterpiracy campaign of the U.S. Navy's West Indies Squadron in the 1820s was the nation and the navy's first long-term challenge following the War of 1812. While the Second Barbary War, with its brief engagements and threat of bombardment followed by immediate capitulation, was over quickly, the Americans faced a complicated threat in the Caribbean from multiple directions, involving not just naval action but also naval diplomacy and creative operational planning. Like the Quasi-War and the First Barbary War, there was no declaration of war. However, the situation was more challenging than those earlier events, because there was not a nation-state enemy. The campaign was an irregular one, forced to deal with nonstate challengers while always remembering the powerful state interests that shared the same seas and ruled the territory from which the pirates operated.

When it came to addressing this irregular threat, the U.S. Navy profited from the lessons of the previous two decades of naval irregular warfare and maritime security operations. From the founding of the West Indies Squadron onward, it was clear to squadron commodores that their junior officers were the key commanders in the success of a campaign. Operating just offshore in small combatants or inside the reefs in cutters and barges, the lieutenants not only commanded the actual combat operations but also conducted their own naval diplomacy and interaction with local authorities. While Biddle, Porter, and Warrington did occasionally patrol, for the most part they focused their own actions on command and control as well as the vital requirements of diplomacy. This was especially the case with the essentially diplomatic South American missions of Oliver Hazard Perry and Charles Morris.

Lt. Francis Gregory brought his experience in small boats and irregular warfare, discussed in chapter 3, and his experience on New Orleans Station prior to the War of 1812, to his successful command of *Grampus* for Biddle and Porter. Lawrence Kearny's multiple captures while in command of *Enterprise* were preceded by similar experience in the war, where he commanded the schooner *Caroline* on patrols of the southern coast and led small-boat raids on British parties. Isaac McKeever, who commanded *Sea Gull* and the combined operation with *Dartmouth*'s boats, had served with Daniel Patterson at New Orleans, working the Gulf Coast during the war and commanding a gunboat in the Battle of Lake Borgne. These lieutenants, and many of their peers, demonstrated the value of command experience at a junior rank in irregular operations. All continued to rise through the ranks. Gregory became one of the navy's first admirals, recalled to duty during the Civil War to oversee the creation of the blockading squadrons. Kearny's naval diplomacy as a commodore in China and his nuanced defense of Hawaii against British annexation in the 1840s were central in the next decades of American foreign policy. The scope of this chapter did not allow for the inclusion of every raid or operation conducted in the multiyear campaign in the West Indies, and there were many, but the examples included validate this larger observation. This pattern likewise carried over from the events in the Caribbean to the Mediterranean in the late 1820s and the Falklands in 1832.[105]

The campaign in the West Indies also demonstrated the patterns of platforms and equipment in naval irregular warfare. Where the commodores during the Quasi-War had to take captured ships into American service to get the small vessels they needed, and Edward Preble went through repeated efforts to find a gunboat flotilla for his operations against Tripoli, in the West Indies the navy realized that successful irregular warfare required a task-organized and -equipped force. After only a year of operations under Biddle, the secretary and the navy commissioners designed and proposed a force built around a specific operational concept for counterpiracy operations. The Mosquito Fleet embraced the craft the pirates and privateers used most effectively—schooners built in the Baltimore clipper style and oared launches and barges—and procured a sufficient force of them. Despite the operational success, the insight about balancing forces, which the campaign could have provided for overall fleet design, appeared to be lost.

The Mosquito Fleet barges deteriorated quickly from poor maintenance and hard use, and the squadron abandoned or scuttled most of them in the Florida Keys. The schooners did not last much longer, and once they were sold or broken up, Congress did not see fit to replace them in order to maintain balance in the overall naval force. Just as most of the schooners and brigs were disposed of following the War of 1812, following the West Indies Campaign the navy again refused to acknowledge the importance of the smallest ships, which were most effective in shallow waters and irregular operations.[106]

In addition to this temporary "fighting fire with fire" force construction, Porter and his compatriots on the board made the innovative recommendation of adding a steam vessel to work in the Cuban shallows and pursue the pirates to their camps and safe havens. The history of the transition from sail to steam in the U.S. Navy almost entirely ignores this development.[107] This despite the fact that *Sea Gull* served as a commissioned warship in combat for over two years, and then served until 1840 as a receiving ship at the Philadelphia Naval Station.[108] Greater historical inquiry into the service of the converted ferryboat is necessary, but the surface examination provided in this chapter demonstrates that deployment of the right ships, with sizes and capabilities appropriate for the operational tasks assigned, and a willingness to embrace new technology, remained important elements of success in naval irregular warfare during the period. Following *Sea Gull*'s return to the United States and her mooring as a receiving ship, there was little acknowledgment by naval leaders that steam would be the future of naval ship design. As Howard Chappelle documented, senior officers believed they might serve as harbor vessels but "a steamer would not do as a seagoing man-of-war."[109] It would take another decade for the conventional wisdom to begin to change.

Finally, the operations against piracy and privateering off Cuba and Puerto Rico also demonstrated the centrality of developing and encouraging partnerships, and introduced multinational operations to the history of American naval irregular warfare. An often-overlooked element of partnership is the one between U.S. naval forces and merchants interested in their own protection. American traders, who worked out of Cuban ports like Matanzas and Havana, as well as San Juan and St. Thomas, regularly provided important intelligence to the West Indies Squadron and on occasion even helped obtain shallow-draft vessels to augment the navy's local forces. Examples like Seaman Daniel Collins demonstrate that some of the navy sailors who enlisted for the counterpiracy campaigns had personal experience with sailing the waters of Cuba, and with the violence of piracy. The regular attempts at cooperation with both local and more powerful colonial officials in the Spanish islands had mixed results, but there were certainly successes and incidents when the Americans and Spanish authorities worked well together. Finally, American cooperation with the British forces in the Caribbean marked a turning point in U.S. relations with Great Britain. While the politicians in London and Washington would take almost a century to conclude that Anglo-American interests coincided more often than they diverged, in the 1820s the U.S. Navy and Royal Navy not only understood this reality, but also started to act upon it by conducting combined operations. After the bitter struggle of the War of 1812, both navies understood only too well that the defense of the international trading system at sea was important to their national well-being. Their cooperation off Cuba suggests this was true, despite continued disagreement over the conditions of belligerent and neutral rights and American inconsistency over the slave trade. Perhaps even more significant than the tactical success experienced when the two forces worked together in the Caribbean was the fact that, despite having been at war, twice in the past decades they never again found themselves in combat with one another.

The West Indies counterpiracy campaign marked the beginning of a new role for the United States Navy on the world's oceans. In the aftermath of the War of 1812, and the national embrace of the need for a navy to defend American interests, the operations off Cuba and Puerto Rico ushered in a willingness to use naval irregular operations globally in defense of both American interests and western norms of free trade and diplomacy. Where previous engagement in irregular combat operations occurred in wars between nations, both declared and undeclared, in the remainder of the Age of Sail the U.S. Navy assumed the mantle of defender of American interests globally, not only in diplomacy and showing the flag but also at the sharp tip of the cutlass and barrel's end of the musket. In the Mediterranean, American naval forces combatted piracy in the Greek islands;[110] on the coast of Africa, Matthew Calbraith Perry launched operations against pirates and slavers that included raiding ashore;[111] and as the next two chapters show, American naval irregular warfare spread to the far side of the world with operations on the seam between the Indian and Pacific oceans.

CHAPTER 7

FIRST SUMATRA EXPEDITION
1831–1832

American merchant traffic and commercial interests were a global endeavor going well back to the colonial period. Discussions over the need for a United States navy in the *Federalist Papers* and debates leading up to the drafting of the Constitution were securely founded in the need to protect American commerce on the high seas, both close to the nation's shores and far from them. In 1790, the trading vessel *Columbia* returned to Boston, the first American merchant ship to circumnavigate the globe. The reports of Capt. Robert Gray of *Columbia*, close on the heels of the 1784 trip of the New York ship *Empress of China* to Canton, set New Englanders' sights toward the Pacific. Whalers, fur traders, and sealers all began regular voyages far from American shores, to new seas where the U.S. Navy would need to sail to protect Americans and American interests.[1]

Captains and shipowners from Salem, Massachusetts, focused on one specific cargo from Southeast Asia: pepper. These merchants founded the pepper and spice trade between Sumatra and the United States. The brig *Rajah*, under Capt. Jonathan Carnes, returned with the first cargo from the coast of Sumatra in 1799, making Carnes and owners Jonathan and Willard Peele and Ebenezr Beckford a small fortune. The small town of Salem took on a manic focus on the spice trade, which would last through the 1850s. The Salem men came to dominate the trade in Southeast Asian waters, which were well known for their riches but also for maritime insecurity.[2]

On 20 June 1831, Capt. John Downes received orders from Secretary of the Navy Levi Woodbury for a cruise to the Pacific. Downes was elevated to the position of commodore and given command of the squadron on the U.S. Navy's Pacific Station, responsible for three ships. The secretary assigned the frigate *Potomac* as his flagship, which was quay-side in Norfolk, Virginia. On the far side of the world he would join the sloop *Falmouth* and schooner *Dolphin*, which were already at sea. En route his orders included the transportation of Martin Van Buren to Portsmouth, England. President Andrew Jackson's administration was dispatching the future president to London to assume his position as the United States' minister to England. Woodbury emphasized to Downes, "[Y]ou will on all occasions render, to our citizens, vessels, commerce, and interests, that assistance and protection to which they are lawfully entitled."[3]

Commodore Downes began his naval career at the age of sixteen when he joined his father on board *Constitution*. Some of the men who rose to prominence in the U.S. Navy had fathers who were officers in either the Continental Navy or early U.S. Navy. But unlike his friend and mentor David Porter, or his future commander Stephen Decatur, Downes came from the other end of the chain of command. His father served as the steward to *Constitution*'s purser. When young John went to sea for the first time in 1800 it was as a waiter working for his father and the other stewards on the ship.[4]

In 1802, he received his coveted appointment as a midshipman and sailed to the Barbary Coast on board *New York*, under the command of Capt. James Barron. The ship sailed as part of the first squadron sent in response to Tripoli's declaration of war. When *New York* arrived at Gibraltar, Commo. Richard Morris took the ship as his flagship. It was during this cruise that Midshipman Downes first met David Porter, who was serving as first lieutenant on board *New York*. In early June 1803, Porter led an expedition of small boats from *New York*, *John Adams*, and *Enterprise* against a convoy of small merchant boats attempting to break the blockade of Tripoli harbor. Once they realized they had been discovered the blockade runners tried to land their cargo in a small bay short of the harbor. In the ensuing landing of sailors and marines, the Americans fought a detachment of Tripolitan soldiers and burned the merchant craft before retreating under the covering fire of the frigates. Downes was one of three midshipmen Porter took with him to command the boats from *New York*.[5]

Downes made an impression on Porter. They parted ways when *New York* returned home later that summer and Porter transferred to *Philadelphia* to remain in the Mediterranean for the war. Following the Barbary War, Downes was promoted to lieutenant in 1807 and two years later, when recently promoted Captain Porter took command of *Essex*, he asked his former raiding partner to join him as his first lieutenant. They served together on *Essex*'s raiding campaign in the Pacific during the War of 1812, and then Downes served with Decatur in the Mediterranean. His promotion to the largely honorary rank of commodore, and command of the Pacific Squadron, was the pinnacle of a busy naval career.[6]

Of the three ships that Downes would command in the Pacific squadron, *Falmouth* and *Dolphin* were the smaller. *Falmouth* was under the command of Master Commandant Francis Gregory and was already in the Pacific patrolling and protecting the American whaling fleet. *Dolphin*, under the command of Lt. John C. Long, had spent the vast majority of its service life, since its launch in

1821, in the Pacific and was already there as well. *Potomac*, Downes's flagship, was the U.S. Navy's newest frigate. Built to the lines of the Raritan Class, she was 1,726 tons, 177 feet long, rated at 44 guns (but carried over 50), and had a crew of 480 sailors and marines. In Norfolk, she was fitting out and planned to transfer to New York to complete the on-load of stores and pick up the diplomatic passenger for the Atlantic crossing.[7]

In the late 1820s and early 1830s, American trade in East Asia was growing, and President Andrew Jackson and his administration recognized the potential of these new markets and resources. Under his administration, the United States began to establish its first trade agreements and treaties with powers in the region.[8] However, Jackson and his advisors also realized if the United States was to have a real presence in the Pacific it would have to demonstrate the ability to protect its citizens and merchants. The three-ship squadron, and in particular the deployment of a large warship like *Potomac* as flagship, was a significant increase over the single cruising naval vessels that the Navy Department had deployed in the past.[9] Woodbury wrote that he hoped the three ships, making up a balanced squadron with the heavy guns of *Potomac* mixed with the shallower draft of the small ships, "will be competent to afford sufficient protection to our extensive and important interests in that region of the world."[10]

Downes's orders were as much diplomatic as they were military. Secretary Woodbury instructed him to stop in the Cape Verde Islands, San Salvador, and Rio de Janeiro on his way to the Pacific to consult with American commercial agents and diplomatic consuls, and ensure they had no pressing needs for security or protection of American interests. He received copies of the treaties between the United States and foreign powers, as well as regulations governing the relationships between naval commanders and American diplomats abroad. Woodbury ensured that Downes understood the generally friendly international relationships he could expect in the region, but also allowed the possibility of trouble either on the west coast of South America or from "freebooters." Woodbury reminded Downes that "cases may arise which it is impossible to foresee, and to meet which definite instructions cannot be given; should such occur, out of the ordinary way, you must be left to the exercise of a sound discretion."[11] Thus instructed, Downes set about preparing his ship for sea.

Less than a month later, world events changed the plan for Downes and *Potomac*. On 16 July, a spice trader named *Friendship* rounded Cape Cod, returning from a voyage to the South China Sea, which had lasted over a year. The ship brought back a dramatic story of piracy, murder, and battle on the island of Sumatra. On 20 July, the owners of *Friendship* wrote to President Jackson outlining the events, which occurred in February 1831, when a band of Acehnese from the village of Kuala Batu attacked their ship, killed three men, and stole thousands of dollars in cash and goods.[12]

The Navy Department was already taking action by the time the president received the letter on 23 July and passed it on to Secretary Levi Woodbury's office. A few days prior, newspaper stories had appeared outlining the attack on *Friendship* and word spread up and down the American seacoast. Based on newspaper reports, Secretary Woodbury sent a letter to Nathaniel Silsbee, one of the ship's owners, on 22 July to gather further information. Silsbee was serving as one of Massachusetts's senators. He and Woodbury were acquaintances in Washington, fellow New Englanders, and the secretary wanted to ensure that a member of Congress knew the navy was ready to assist. When the president officially referred the matter to him the next day, Woodbury immediately sent another letter to all three owners specifically requesting detailed affidavits relating to the attack, as well as information on the locale where it occurred.[13] The owners replied with excerpts from *Friendship*'s logbook and statements of the crew.[14]

In February 1831, the American merchant sloop *Friendship* had sailed into the small harbor at Kuala Batu, Sumatra, to purchase a load of pepper. The village was located on the northwestern shore of the island, along a section of the coast with several villages regularly visited by American and European traders. It was early in the season and Capt. Charles Endicott and his crew were the first Americans on the coast to start negotiating for a cargo. They had missed the opportunity at the end of the 1830 pepper season, arriving from the United States too late to purchase enough pepper to fill their hold. Rather than return home without a full cargo, Endicott elected to stay in the Pacific to wait for the start of the 1831 season.[15]

Friendship prepared to take on her cargo, and Captain Endicott went ashore on 7 February with his second officer, John Barry, and four sailors to meet the local pepper merchants at the village's pepper scales. They worked all morning and into the afternoon to fill one of the village's boats with a load of pepper. Around three in the afternoon the boat headed down Stone River for the harbor, where locals would transfer the bagged pepper to the crew on board *Friendship*.[16] When the boat arrived alongside, however, it carried not only the bags of pepper and oarsmen, but also approximately half a dozen extra men. The men boarded the merchant ship, against the orders Endicott had left behind. When Chief Mate Charles Knight, a Salem native and experienced hand in the waters of Southeast Asia, realized he had local men on deck he halted the loading and ordered them off the ship. The Acehnese outnumbered the Americans on deck and immediately attacked. Knight and members of the crew fought back but were overwhelmed. As defense became hopeless, with Knight and two other members of the crew already dead, a group of men leapt overboard and swam for the far shore, away from the village.[17]

John Barry, *Friendship*'s second officer, and two members of the party ashore saw the commotion on board the ship and watched as their shipmates dove overboard. They hurried from the beach to report to Endicott, who later related he was "convinced from this circumstance that we on shore had no time to lose, we immediately sprang into the ship's boat, and pushed off." They were joined by a local Acehnese, known to the Americans as Po Adam, who regularly served as a guide and liaison for American pepper traders along that section of coast.[18] Local boats gave chase, but the Americans were able to clear the harbor and headed for the nearest trading village. They arrived off of Muckie after midnight and found the American ships *James Monroe* of New York, *Governor Endicott* of Salem, and *Palmer* of Boston riding at anchor.[19] At sunrise *James Monroe*, *Governor Endicott*, and *Palmer* sailed from Muckie to retake *Friendship*. The next morning, after the local rajah rebuffed their emissary, the three ships opened fire on the pirates and the village's forts, as Endicott manned a boarding party and led three boats full of armed men to retake his ship. They found it deserted, and with the other three ships sailed clear of the harbor.[20]

As the details of the attack became clear in Washington, Secretary Woodbury drew up a new set of orders for Commodore Downes on his trip to the Pacific Station. In the original orders, the secretary had addressed the subject of piracy. However, he was more concerned that "freebooters," who had been cleared from the Caribbean during the counterpiracy campaign led by Porter and Warrington, might reappear on the west coast of Mexico and begin attacks in the Pacific. Operations of the U.S. Navy's Pacific Squadron focused primarily on the west coast of North and South America at the time, in the eastern half of the Pacific Ocean. The navy and its officers gave little thought to dealing with problems in archipelagic Southeast Asia.[21]

The attack on *Friendship* caught the Navy Department and the Jackson administration by surprise. It did not surprise the Salem merchants who traded on the Sumatran coast. Local pirates attacked a number of ships over the years, including the massacre of a part of the crew of *Putnam* in 1805, and an attack on *Marquis de Somerulas* in late 1807. The owners of *Friendship* emphasized in their letter to President Jackson that similar attacks on British merchants in the region had resulted in the dispatch of a warship and retribution for the offenders. That was why, they claimed, British merchants traded in the waters of the Indian Ocean and South China Sea with little fear for their safety.[22] In the early decades of American trade, there had been limited U.S. Navy presence in Southeast Asia. In 1800, Edward Preble sailed *Essex* to the region to convoy merchantmen home during the Quasi-War, which was the first American naval voyage to cross the equator, never mind leave the Atlantic.[23] From the Quasi-War to the sailing of *Potomac*, only three other ships sailed to Southeast Asia. The most recent had been in 1826 when the sloop of war *Vincennes*, under the command of Master Commandant William Bolton, completed the U.S. Navy's first circumnavigation of the globe. Besides those two voyages, local inhabitants had never seen American warships in the waters of the South China Sea and eastern Indian Ocean.[24]

In receipt of Silsbee's letter, Woodbury wrote that same day to Commodore Downes and apprised him that his mission had been changed. *Potomac* was to sail as soon as possible for the Pacific to investigate and avenge the "outrage." He promised the commodore anything he needed in support but warned him to prepare for sea as rapidly as possible. Before sailing, he would receive a new set of official orders along with detailed accounts of the attack on *Friendship*, and what he should expect along the coast of Sumatra. The department had already begun a search for a sailing master with personal knowledge of the region. During the last days of July and first week of August, Woodbury and Downes kept up a steady correspondence as the secretary moved resources from Philadelphia and Boston toward New York to fill the commodore's needs. On 8 August, Woodbury wrote to say that John Barry, the former second officer of *Friendship*, was en route from Salem. On his arrival at New York, he received a warrant signed by the secretary, appointing him as a sailing master in the U.S. Navy to accompany *Potomac*.[25]

On 9 August 1831, John Downes received new official orders for his deployment to the Pacific. He was "directed to repair at once to Sumatra" and given explicit instructions on how he was to engage with the locals. Upon arrival on the coast, he was ordered to obtain as much information as he could about the attack on *Friendship* from Americans trading in those waters. If it appeared the events were in agreement with the story related by the crew, the president ordered Downes to sail for Kuala Batu. There he was to demand the rajah provide restitution for the material loss, and immediate punishment of the men responsible for the murder of the three Americans. If the rajah refused, Downes was to land a force, take custody of the responsible parties himself, and "destroy the boats and vessels of any kind engaged in the piracy, and the forts and dwellings near the scene of aggression, used for shelter and defense; and to give public information to the population there collected, that if full restitution is not speedily made, and forbearance exercised hereafter from the like piracies and murders upon American citizens, other ships-of-war will soon be dispatched thither, to inflict a more ample punishment."[26] Downes was to return the offenders to the United States for trial, by whatever method he found convenient, and proceed to assume his duties in command on Pacific Station.

On 24 August 1831, USS *Potomac* put to sea. She carried more than a full complement of officers and men, with a total of five hundred sailors and marines on board, and Sailing Master John Barry, who had arrived just a week before. Secretary Woodbury had rushed men and officers from shore stations around the eastern seaboard to ensure Downes had a full crew. The commodore's mandate seemed clear.[27]

Potomac reached Rio de Janeiro on 16 October, anchoring in the harbor the morning after a fifty-one-day passage.[28] Downes and his crew joined USS *Lexington*, a sloop of war, as well as warships from Great Britain, Sweden, and France in the harbor. Because of a diplomatic row between the British and Brazilian governments, and a large number of ships in port, it took some time for all the wood, water, and stores ordered by *Potomac* to arrive. It was almost three weeks before the ship was ready again for sea.[29] On the morning of 5 November, the frigate weighed anchor. Light and variable winds made departure difficult and Downes ordered the ship's boats in the water to tow her clear of the harbor. A number of other international ships at anchor sent their boats to assist as well, and by late afternoon *Potomac* had cleared the shoals and set sail for the Cape of Good Hope.[30]

After a thirty-one-day gale-filled passage, *Potomac* arrived at Table Bay, Cape Town, on 7 December and exchanged salutes with the harbor fortress.[31] The crew set about making repairs to the ship and loading stores for the long passage around the Cape and across the Indian Ocean. On the night after they arrived, the British governor Sir Lowry Cole invited Downes and his officers to dine with him. There, the commodore met a number of merchants and British officers who had spent time in India and in the islands of Southeast Asia. The following night he accepted the invitation of the colonel commanding the 72nd Regiment of Scottish Highlanders, who manned the garrison and fortress overlooking the harbor. Some of the men had spent time in India as well. During both dinners, Downes took the opportunity to gather information about the inhabitants of Sumatra.[32]

The information Downes obtained from the British appeared to confirm his impression from the information passed by Woodbury. The documents forwarded from Salem suggested that the locals of Kuala Batu and the northwest coast of Sumatra were of a suspicious nature and untrustworthy. The villages of the region notionally came under the sovereignty of the king of the Acehnese, who ruled over the northern part of the island while the Dutch laid claim to the south. However, these political connections were loose, the court claiming no responsibility for actions of the local rajahs. Open conflict between villages was very common and it was not unusual for them to "wantonly plunder and kill strangers." According to Salem merchants who provided the information to the Department of the Navy, corruption and violence against Americans had increased in recent years.[33]

The consensus in Cape Town was along similar lines. Officers and merchants Downes spoke with assured him that the men of the Sumatran coast were known for their treachery and piracy. They also warned the Americans not to underestimate the locals. The British told Downes they were courageous and skilled fighters who were also crafty. Downes was left with the impression that he could leave nothing to chance, or the Acehnese would take advantage of him. Officers of both the British Army and Royal Navy, many who claimed to have been on the northwest coast of Sumatra, warned him that the inhabitants "were cruel and treacherous on all parts of the coast, and by no means to be trusted."[34] He was also told to take care when displaying his intentions, because the inhabitants of Kuala Batu were likely to pick up their valuables and disappear into the jungle if he gave them enough warning. The officers related to him the story of a recent British expedition against another village. They arrived in force, not disguising their intentions. When they landed, they discovered the village empty of all inhabitants and valuables. However, once they let their guard down local forces attacked the British party on the beach as they attempted to embark their boats and return to their ships. The counterattack slaughtered the landing force.[35]

Potomac set sail from Table Bay on 12 December and headed into the Indian Ocean while the seriousness of the mission began to register with the officers and men. Downes called all hands to quarters once clear of land and headed east. Francis Warriner, the ship's schoolmaster, recorded: "Commodore Downes addressed the crew on the magnitude and importance of the enterprise on which they were sent, and the absolute necessity that every man should thoroughly understand and faithfully discharge his duty." The intelligence gathered at Cape Town, documents sent by the secretary of the navy, information gleaned during the ship's stay in Rio, and from Sailing Master Barry, convinced Downes that he could leave nothing to chance on his ship's mission to Sumatra.[36]

The commodore turned to his officers to help develop a plan for the arrival off of Sumatra. Based on the advice that the local villagers were likely to disappear into the jungle if given warning, they came up with a plan for landing a large force, which could surround the small forts guarding the village. With the strength of the large American force evident, and escape impossible, Downes would then open dialogue with the rajahs and make demands for justice and restitution. The officers assigned half of *Potomac*'s crew, 250 men, to the landing party, and the other half would remain behind to protect the ship and man the guns to provide fire support to the landing party. Sailing Master Barry would pilot the boats, and guide the force one ashore, because of his direct knowledge of the layout of the shoreline and village. The commodore assigned overall command of the expedition to Lt. Irvine Shubrick, *Potomac*'s first lieutenant.

The landing force was broken up into four divisions. The first to land would be the division of *Potomac*'s marines, armed with their muskets and pistols. The three divisions of sailors would follow, each commanded by one of the ship's lieutenants, assisted by three midshipmen. The sailors armed themselves with an assortment of pikes, cutlasses, pistols and muskets. The plan also assigned one of the naval divisions a small 6-pound artillery piece, which Commodore Downes had obtained in New York before sailing. To accompany the landing-force boats, the crew outfitted the ship's launch with a 12-pound carronade in the bow and swivel guns on each side. One of the midshipmen received command of the "gunboat," which would offer fire support close to the beach.[37]

With a general outline of the operation prepared, the officers and marines developed a plan to train the crew. Schoolmaster Warriner recorded in his journal that, as was the custom on most naval ships, "the sailors had been occasionally drilled at the small arms, and were disposed to laugh at the whole affair."[38] As *Potomac* left Africa behind her, things started to change. The men assigned to the landing party began to conduct daily drill on deck. Training by division, the ship's marines helped teach the men what they would need to know for operations ashore. Along with the divisional training, at least twice a week the entire landing force rehearsed a part of the plan.[39] J. N. Reynolds, Downes's personal secretary, wrote in his record of the deployment, "[T]hese exercises, and the object to which they led, seemed now to engross the minds and feelings of the whole crew; so that the ordinary tediousness of headwinds and cross seas was but partially felt." Assistant Surgeon J. M. Foltz wrote in his medical summary that during the passage to Sumatra the sick list was the shortest length he saw during the entirety of the three-year deployment. In his report, he attributed the apparent health of the crew more to the excitement over the approaching mission than the medical reality on board.[40]

As the ship approached Sumatra, Commodore Downes ordered the crew to begin changing the rigging of the ship. In his desire not to set off an alarm along the coast, he later wrote to Secretary Woodbury that he had "taken care to disguise" the ship and make her look like a large East Indiaman.[41] The sailors had run in the guns on the main deck, and then arranged them in the center of the deck and covered them with spare sailcloth and hammocks. The crew closed the gun ports, painting over some of them so from a distance it appeared the ship had only ten guns per side. They also hauled up stump top-gallant masts, which made the ship's rigging appear smaller and look closer to how merchant vessels, with smaller crews, would have been rigged.[42]

At five in the morning, on 3 February 1832, Commodore Downes had the crew beat to quarters for inspection. An hour later, the lookouts called out the shoreline of Sumatra. Taking regular soundings and proceeding slowly, *Potomac* picked her way south along the coast. Late in the day, the ship came to anchor several miles from Kuala Batu.[43] The crew spent 4 February preparing for the impending mission. The lookouts kept a sharp eye on the horizon but did not see any American vessels along the coast. A proa was spotted outbound from Kuala Batu and appeared headed in the direction of the ship. Lieutenant Shubrick and Sailing Master Barry took a party in *Potomac*'s whale boat to investigate. As they approached the local craft, they hid their uniform jackets and muskets under a sail in the bottom of the boat. Barry recognized a number of the locals on board the proa, but it did not appear they recognized him. The Americans asked where the boat was bound, and the reply was that they were carrying the rajah's annual tribute to the king. Keeping up their disguise as merchants, the Americans asked if there was pepper available at Kuala Batu. The local men told them there was plenty. The two boats separated, and the local craft continued down the coast. Returning to *Potomac*, Lieutenant Shubrick reported the intelligence gained and the men confirmed the ship looked more like a merchant than a warship. *Potomac* spent the night at anchor.[44] Downes wrote in a letter to Woodbury after the events at Kuala Batu, "[F]inding no vessel on the coast, I could obtain no information, in addition to that already possessed."[45]

After noon on 5 February, the crew weighed anchor. With the nearby trading villages Muckie, Soo Soo, and Kuala Batu in sight, *Potomac* made her way toward the harbor at Kuala Batu. The lookouts spotted a number of small sail from local craft working the shoreline, and one western merchant flying British colors, but Downes did not hail any of them. Approaching the harbor, *Potomac* raised a Danish flag in order to avoid suspicion of their true purpose and most of the men hid below decks. Schoolmaster Warriner recorded in his journal: "[E]very port being closed, the air that we breathed was close and stifled. The melted tar fell in drops upon the deck, and fairly boiled from the seams between the planks."[46]

Anchoring directly off of Kuala Batu, Commodore Downes sent a party, again under the command of Lieutenant Shubrick, toward the beach to gather intelligence. Dressed in civilian clothes, Shubrick posed as a ship's officer and marine first lieutenant Alvin Edson played the role of supercargo. Barry briefed the party on how normal pepper merchants behaved, and the process for negotiating the price for a cargo. The boat crossed several miles of open water toward the village, and as they approached men began to gather on the beach. Upward of two hundred local men, armed with blunderbusses, knives, and *krisses* (a local long-bladed weapon) met the boat as it approached the surf line. Shubrick elected not to land, based on both the reception party and the rising surf. He did initiate a shouted dialogue from the boat over the price of pepper and began some bargaining with the locals. However, the rajah would not come down to the beach, and the lieutenant told the locals the captain of the ship would return the next day to finish the negotiation on shore. The party rowed back to the ship.[47]

The "inconclusive result" of the reconnaissance, as Downes would characterize it, left the commodore without an easy answer for how to proceed.[48] The quick gathering of armed men on the beach, in response to a single boat, convinced him that landing during the day with a large party would have one of two possible results. The Acehnese would either flee into the jungle, or they would mass and attack the landing force as it attempted to make it through the surf, the most dangerous part of any amphibious operation. The commodore concluded that his original plan, of a large landing conducted in secret and meant to surround the village before opening negotiations with local authorities, was the only option with a chance for success. He ordered his men to begin preparing the landing force. He wrote, "[N]o demand of satisfaction was made previous to my [landing], because I was satisfied, from what knowledge I had already of the character of the people, that no such demand would be answered, except only by refusal." The muskets were cleaned, equipment prepared, and cutlasses sharpened.[49]

Downes gathered his wardroom and reviewed the plan with Shubrick and the other officers. Their purpose was to land and surround the fortresses, with the goal of keeping any of the perpetrators of the attack on *Friendship* from escaping. The landing was set for the dark hours before dawn, so when the village awoke they would discover themselves surrounded and unable to resist. Downes emphasized that the village was to be given the opportunity to give up the attackers, but if no action was forthcoming the landing force would attack, "but were directed to spare the women and children."[50]

John Barry went over the layout of the village with the officers and they reviewed their plan. At the center of the village was the bazaar and smokehouses, with dwellings arranged just to the northwest of them with the village mosque. Along the small and narrow Stone River, there were boatsheds, more smoke-houses, and the pepper scales. Around the perimeter of the village were five primitive fortresses of bamboo stockade fences, with elevated firing platforms for small cannon and swivel guns. Four of them were located on the northwest side of the river and the fifth on the opposite bank. Each of the five most powerful leaders in the village, the rajahs, controlled one of the forts. Shubrick assigned one division to each of the four forts on the northwest side of the river, with orders to first surround it so nobody could escape, but then, if necessary and on the orders of Lieutenant Shubrick, to reduce it.[51]

Under cover of darkness, the men left their hiding places on the lower decks and prepared the boats and equipment for the landing. After midnight, the crews lowered the boats to the water on the seaward side of *Potomac* and the men mustered on deck in their boat teams. It took a little over two hours to embark the entire force of 282 men who went ashore, which was longer than they had planned. At 0215 the boats cast off for the designated landing beach, located a mile and a half north of the village itself.[52] The four divisions made their landing under a moonless night sky, in heavy surf. The division of marines put out pickets and mustered on the beach, forming the head of the column. In the high surf, the landing was a dangerous evolution but the boat crews completed it without incident. Once the sailors were on the beach, it was a different story than with the marines. Less experienced at military movement, it took more time to get them into formation behind the marine division for the march to the village. Some of the marines served as what today's amphibious forces call beach masters: helping the sailors find their divisions, mustering, and forming them up. Shubrick wrote in his report after the attack, "I feel much indebted to [marine] Lieutenants Edson and Terrett for the promptness displayed by them in forming the Marines, and in assisting and forming the other divisions: all of which was effected, with coolness."[53]

The boats, under the direction of Passed Midn. Sylvanus W. Godon, moved back off the beach. As the sun began to rise over the mountains, the formation set off from the beach toward the village. When the column approached the village, the lead men spotted a figure dashing off through the underbrush toward the town. The officers realized their element of surprise was about to be lost. Approaching the first fort, Lt. Henry Hoff and his division broke away from the column toward their assigned target. As they advanced defenders opened fire, and the opportunity for the Americans to surround and overawe the village evaporated. The battle commenced at 0515, and on board *Potomac* the officer of the watch immediately hauled down the Danish colors and raised the American flag.[54]

Hoff organized his men to surround the fort and return fire, establishing a siege of the forces inside. Lieutenant Pinkham and his naval division, and Edson with his marines, moved around the rear of the village to attack their assigned forts. Shubrick led the final naval division, with Lieutenant Ingersoll and the artillery piece, along the beach toward the largest of the forts, located near the harbor's edge. On board the ship, the crew hung in the rigging or at their guns, watching their shipmates locked in combat.[55] Shubrick, with Ingersoll's division, approached the largest fort and ordered the artillery placed approximately fifty yards from the enemy's walls and then opened fire. Pinkham's division, unable to locate their assigned target at the rear of the village despite having Sailing Master

Barry along as a guide, joined the assault on the main fort.[56] Hoff's pioneers broke through the palisade wall surrounding the first fort and drove off the defenders. The rajah, known to the Americans as Po Mahomet and identified by the name To'onkou N'Yamat in Warriner's journal, who by all accounts was the village leader most responsible for the attack on *Friendship*, owned the fort. The sailors and marines killed him in the assault.[57]

The first American death occurred on the inner wall of the large central fort, as Seaman William Smith attempted to climb it and a defender shot him through the head. Two other men were injured, and Shubrick ordered the men to pull back. The gunboat under Midshipman Godon opened fire on the front of the fort, joining the artillery piece. Shubrick ordered the outer section of the fort set on fire, along with the nearby buildings that made up part of the bazaar. His hope was the fire would spread to the inner section of the fort, but with no wind to spread the flames, the buildings burned without the fire spreading fast enough to affect the battle.[58] The marines under Edson forced the defenders in the fort at the rear of the village to surrender. Edson left Second Lieutenant Terrett with a small detachment in the fort to hold it, and led the remainder of his marines back through the center of the village to join the two divisions already massed around the harbor front fort. Lieutenant Shubrick reported after the battle: "[S]o tenaciously did the enemy cling to their positions, that not until nearly all of them had been destroyed could we carry the fort. . . . The American colours were then hoisted, with three hearty cheers [*sic*]."[59]

Inside the fort, the Americans discovered a trail of gunpowder leading to the inner magazine. Fearing there might be other traps set, Shubrick ordered his men out of the wrecked structure.[60] The battle degenerated into a general action, as the flames began to spread among the village's dwellings and sailors and marines fought the remaining Acehnese back into the jungle. Warriner wrote in his journal, after talking with men who had been on the ground, "[M]en, women, and children were seen flying in every direction, carrying the few articles they were able to seize in their moment of peril, and some of the men were cut down in their flight."[61] The marines attempted to maintain order, but sailors began looking for plunder, and as the surf began to rise, Lieutenant Shubrick ordered the recall.[62]

Boats began taking the dead and wounded back to the ship. As the men assembled to embark for the return to *Potomac*, the marines established a defensive perimeter and sailors loaded the boats. When the first group of boats departed the beach, they took a brief cannonading from the southeast side of Stone River. The landing force had not assaulted the fort on the other side of the river. Fighting their way across the river was more trouble than it was worth, as the village of Kuala Batu burned around the Americans. The brief fire halted, however, and by ten o'clock the whole of the landing force had safely returned to the ship, greeted by a relieved Commodore Downes.

The men of *Potomac* spent the rest of the day treating the wounded and getting the ship back into her normal condition. The bodies of Seaman Smith and Pvt. Benjamin Brown, killed in the fighting at the back end of the village, were first to be returned to the ship. Eleven wounded men followed, including First Lieutenant Edson and two of his marines, one of them wounded severely, and Midn. J. W. Taylor, who had been part of the assault on the walls of the major fort. Assistant Surgeons Foltz and H. D. Pawling worked through the day to stabilize their patients. At 1100, the chaplain and crew held a service for Smith and Brown and their bodies were committed to the deep. The guns, which the crew had relocated to disguise the ship, were remounted on the gun deck, gun ports open, and *Potomac*'s massive broadside was run out. The sailors returned the ship to full fighting trim.[63]

Lieutenant Shubrick estimated that over 150 Sumatrans died in the fighting, including "Po Mahomet, the principal rajah concerned in the plunder and massacre of the crew of the ship friendship [sic]."[64] An uncounted number of Sumatran casualties were villagers, including women and children. Many locals escaped into the surrounding jungle, both civilians and fighters, and while Shubrick's count is the only source available it is likely inaccurate. The Americans captured a number of small arms and one brass field piece, and they spiked all the cannon found in the forts. The fires set burned the majority of the village, as well as a number of the boat-building stocks, proas, and other small boats along the beach.[65]

The next morning, the anniversary of the original attack on *Friendship*, brought warm and clear weather with light winds out of the south. Lookouts sighted four men in a canoe approaching *Potomac*, but they did not appear to be a threat. Arriving alongside, one of the men introduced himself as Po Adam. He was the local trader who had served as an intermediary for *Friendship* the year before, and who had helped the Americans escape the initial attack. At first, he did not see Barry, but as soon as he did there was a great reunion between the two men. According to Po Adam and his three servants, the king of Aceh had punished him and had reduced his standing in the community, and he incurred the disdain and occasional wrath of the people of Kuala Batu. When *Potomac* arrived, under Danish colors, he had paid the ship no mind. However, as the sounds of battle reached his home, located between Kuala Batu and the village of Soo Soo, he was happy to discover the Americans had returned. Once the Americans had razed the village, he decided to visit the ship. He offered his services as an interpreter and intermediary along the coast if Downes was in need of them.[66]

Downes talked with Barry and Po Adam, who explained that the locals had long doubted the United States had a navy, or at the least one with powerful ships mounting heavy cannon like the British. During his years trading in the region, Barry had warned the rajahs, but conversations appeared to result in the question: if you have a navy, why have we not we seen it? The fort on the southeast side of the river remained armed, and had fired on the embarking landing force. Downes elected to use it to make a demonstration of American firepower.[67]

The commodore ordered the ship to quarters and as the sun approached noon *Potomac* moved in closer to the fort on the opposite side of the Stone River. They dropped anchor a mile off the shore at 1215 and prepared for bombardment. At 1220, Downes ordered the guns to open fire, and the crew kept up a steady but slow fire for an hour. Ten minutes after the firing ceased, white flags rose over the tops of the forts. The Americans saw villagers from neighboring settlements up and down the coast watching the spectacle of destruction unfold. *Potomac* raised a white flag up the main halyard in response to the flags over the forts, and then received a small delegation that came out to the ship in a canoe under a white flag. With Po Adam serving as interpreter, the three men of Kuala Batu confessed the village's involvement with piracy and asked for forgiveness and mercy.[68] Downes reported to Secretary Woodbury that the leader of the delegation "begged that I would grant them peace." He explained to the men the reason that leaders in America sent *Potomac* to their village and told them he was satisfied that his mission to bring the attackers of the *Friendship* to justice was complete. He agreed to maintain peace with Kuala Batu and "at the same time, I assured him that if forbearance should not be exercised thereafter from committing piracies and murders upon American citizens, other ships-of-war would be dispatched to inflict upon them further punishment."[69]

At sunrise on the morning of 8 February, *Potomac* was under way for the short sail southeast along the coast to the village of Soo Soo. Lieutenant Wilson and Sailing Master Barry went ashore along with Midshipman Lincoln and a guard of marines, to arrange to obtain water. Wilson saw white flags all down the coast, and carried one in the boat with him. The local villagers claimed the day was a holiday, and they themselves would not work, but they allowed the crew to obtain water themselves. Downes sent the ship's launch and one of the cutters to the beach to collect fresh water in the river alongside the village. The launch, still armed as a gunboat with the carronade and swivels, maintained a watch over the working party and landed a small guard of marines. A number of armed locals gathered to watch the men work but remained outside the marines' perimeter, and the sailors completed the watering without incident.[70]

While *Potomac* was at anchor at Soo Soo, Po Adam alternated spending time on board and time ashore, serving as a local agent for *Potomac* and Commodore Downes with the rajahs. He helped obtain all the stores the ship required, including vegetables, fowl, and several water buffalo the crew considered to be as close to cattle as they could get. He also arranged with local merchants who took sampans out to the ship carrying coconuts, pumpkins, fruit, and sugarcane to trade with the sailors.[71] Delegations from Soo Soo and a number of other villages along the northwest coast of Aceh visited the ship; Po Adam and a Dutch sailor fluent in Malay, who had shipped with *Potomac* at Cape Town, served as interpreters.[72] Downes reported to Woodbury, "[A]ll of them have declared their friendly disposition towards the Americans and their desire to obtain our friendship. Corresponding assurances were given on my part and they left the ship apparently well satisfied."[73] The commodore gave the delegations tours of the ship, occasionally firing a cannon seaward to demonstrate the power of the guns, and talked with them. However, he made no attempts at treaties or any formal agreements.[74]

After daily trips to the river, *Potomac* finished watering on 14 February, with 40,000 gallons on board. The ship spent a few more days receiving delegations and finishing loading food stuffs. At 3 o'clock in the morning on 18 February, *Potomac* sailed from Soo Soo to proceed to the Pacific and execute the rest of her orders. Commodore Downes assumed command of the Pacific Squadron for the next two years. The frigate sailed down the coast of Sumatra to the Sunda Strait, and left Kuala Batu and Sumatra behind her.[75]

First notice of the battle that occurred at Kuala Batu reached the United States in July in the form of a letter to the editors of the *New York Evening Post*. Printed anonymously on 5 July, the letter detailed the attack, including the deaths of civilians and the burning of the village. It also clearly stated that there were no negotiations prior to the landing and assault. Other newspapers reprinted the letter and it set off a political firestorm in Washington, D.C. The summer of 1832 was the middle of a presidential election campaign, and President Jackson's Democratic Party was engaged in a very ugly political battle with the National Republicans, or Whig Party, over the national bank. The prospect of a military action on the other side of the world, and the deaths of women and children, fueled political turmoil. Major newspapers split along party lines as some condemned the attack and opined it demonstrated the president's callousness and disregard for congressional war powers. Others trumpeted the operation as a brave defense of Americans around the world and the sign of a strong executive.[76]

The press back-and-forth was enough to get Congress in motion, and the House of Representatives ordered the president to report on the events to the Foreign

Affairs Committee. The administration agreed, sending a copy of Downes's orders as well as a copy of the report they received from the commodore after the landing. While political opponents wanted to use the accusations of killing women and children against the president, it was clear Jackson (through Secretary Woodbury) had been explicit about negotiating first, followed by military action only if necessary. Since Downes was still on Pacific Station, and would be for at least two more years, there was no opportunity for him to explain himself, and the political forces at work were hesitant to question a well-regarded naval officer in public. The committee elected to file the reports and end the inquiry rather than court-martial a senior officer in absentia. Motions to publish the records were defeated, and the political firestorm died out. The president and his supporters publicly defended Downes and lauded the bravery of the sailors and marines.[77]

While the administration stood behind their Pacific Squadron commander in public and, in his annual message for 1832, President Jackson credited the operation with increasing respect for the American flag and the safety of American citizens, Secretary Woodbury privately rebuked Downes. He complained that the commodore did not explicitly follow orders and negotiate with the local leaders first. Neither the letter printed in the *New York Evening Post* nor Downes's original report to the secretary had explained the planning and events leading up to the attack. When Downes received Woodbury's letter he immediately responded to explain his plan and why the expedition developed in the way it had.[78]

In the records of "Letters Received by the Secretary of the Navy," multiple responses are present, with different dates but nearly the same content, down to exact duplication of the majority of the language in the letters. Downes appears to have understood the displeasure of the administration and was concerned about the accusation that he had exceeded his authority. It is important to remember his friend and mentor David Porter was court-martialed and had resigned the service because of a very similar charge in the 1820s. Downes himself sat on the Porter court-martial. From his repeated efforts to ensure that he explained his side of the story, Downes obviously did not want to suffer the same fate.[79]

In his explanation, Downes laid out his reasoning for how he and his officers had put the plan together, and he attempted to explain that circumstances got in the way of the intended method of execution. He outlined his fear that the responsible parties would slip away into the jungle if they did not handle the operation well. He also related the information he received during his inquiries in Rio and Cape Town, which he had not shared with the secretary in previous correspondence. The way Downes saw it, the fact that the landing force was spotted and taken under fire before they could affect their encirclement of the town meant that he did not have the opportunity to negotiate with the rajahs. He concludes his letter by saying: "I could not believe for a moment that my government had dispatched a vessel of such dimensions, and through seas so dangerous, without attaching to her movement expectations of national importance."[80]

The orders from Woodbury to Downes instructed him to ensure the parties responsible for the attack on *Friendship* were appropriately punished. The administration hoped the local authorities would carry out the punishment, but if not, Downes was explicitly instructed to obtain custody of the individuals in question. This surely weighed heavily in the commodore's mind as he heard the stories in Rio and Cape Town, which warned him the Acehnese were likely to slip away if he played too strong a hand. From the Warriner and Reynolds narratives, it appears Sailing Master Barry, whom Secretary Woodbury had sent specifically to provide local knowledge and guidance, was giving him similar advice. Once the intelligence convinced Downes his plan would need to account for the containment of the village, his decision to obtain the element of surprise appears more reasonable.[81]

Downes and *Potomac* faced another challenge, which contributed to the possibility of losing their quarry. The Indian Ocean crossing was long and depleted most of *Potomac*'s stores. Two facts illustrated this: first was that Downes kept his crew on partial rations to ensure the food lasted, and second was the 40,000 gallons of water that the sailors needed to obtain once the ship reached Soo Soo.[82] The logistical requirements of keeping a large frigate at sea meant that the Americans did not have an unlimited amount of time to remain off of Kuala Batu. Yet, if *Potomac* put in at South Tallapow, or another local port for resupply, word might spread to Kuala Batu that the Americans were back in force this time. They would, once again, risk the perpetrators' escape.

While the plan to surround the village and obtain the element of surprise appears reasonable, the action ashore most certainly degenerated into a full-fledged fight, which neither Woodbury nor Downes appear to have wanted. There are two main factors for this: one that Downes could have controlled, and a second he could not. The landing force's march to the village was late and they were not in position before sunrise, contributing directly to the fact that local defenders spotted them and the local forts initiated the battle. Twice during the amphibious landing, the process took longer than had been expected. This occurred both on the deck of *Potomac* as they embarked the force in the ship's boats, and then again on the beach as the divisions of sailors formed for the march to the village. It was a failure of both planning, by not building in extra time for unexpected issues, and of execution. *Potomac* also faced a challenge of which the Americans were entirely unaware: the rajahs of Kuala Batu had recently entered into a "difficulty of a serious nature" with the leaders in the villages of Soo Soo and Pulo Kio. As a result, the village was prepared for battle, and the forts were reinforced and rearmed, in the weeks before *Potomac*'s arrival. The mass of armed men on the beach when Lieutenant Shubrick took his reconnaissance boat into the harbor was interpreted as a hostility toward westerners, but more likely was the result of the fact that the village was already on a war footing. The American landing force marched into a prepared defense, not a quiet Sumatran trading village. As a result, the battle was far more violent and bloody than anyone expected.[83]

Three factors, which played a role in the successes and failures of American naval irregular warfare throughout the Age of Sail, were once again important during *Potomac's* mission to Kuala Batu. The people involved in the operation, platforms and military equipment available, and partnerships between U.S. naval forces and other interested parties were all in play. Commodore Downes himself, and his own personal experience, certainly played a significant role. As was mentioned previously, he served on landing parties and even led them during *Essex's* operations in the Pacific during the War of 1812.[84] That experience, early in his career, was likely formative and helps explain some of the things he did during the voyage to Sumatra. When David Porter took command of *Essex*, he knew the ship had limitations. It was on the smaller side of the frigate class, and the ship's armament consisted nearly exclusively of short-range carronades. Recognizing those limits, Porter shipped a crew larger than needed and then gave them an unprecedented amount of training in small arms and hand-to-hand combat, in order to create a crew expert in boarding and small-boat operations.[85]

Later in life David Farragut, who was Porter's adopted son and a midshipman on board *Essex*, wrote: "[An *Essex* sailor was always] the best swordsman on board. They had been so thoroughly trained as boarders, that every man was prepared for such an emergency, with his cutlass as sharp as a razor, a dirk made by the ship's armorer from a file, and a pistol."[86] John Downes, as the first lieutenant on board *Essex*, would have been the officer responsible for developing the training plan for the crew and of scheduling the exercises and practice evolutions, which created the experts described by Farragut. Those, combined with his own leadership of operations ashore during the conflict between the *Essex* and a tribe on Nuku Hiva in the Marquesas during the autumn of 1813, were precursors to *Potomac's* aggressive training plan for the landing force. It also influenced his initial foresight to ship a crew larger than the frigate's rating required, and to bring along an artillery piece, which was not standard equipage. Tactical success in naval irregular warfare relied on the leadership of commanders trained, or experienced, in the mission. It was not a "lesser included" part of traditional naval warfare, with skills that would automatically be gained by preparing for squadron engagements or fleet exercises.

Downes's experience on Nuku Hiva, however, also likely contributed to his diplomatic failings during the *Potomac* voyage, and the questions about the mission's success afterward. Following the battle in Kuala Batu, he made many verbal agreements, and hosted ambassadors and emissaries from the local villages. However, he did not sign any diplomatic agreements or send any formal declarations or treaties back to the United States. As the next chapter will show, this was different from how other U.S. naval officers would handle maritime security challenges in the region. It also contributed directly to the political fallout in Washington and Secretary Woodbury's displeasure at his handling of the mission.

Almost twenty years before in the Marquesas, Porter and Downes had discovered a political system that appeared to lack structure or authority, based primarily on familial and tribal ties. Porter wrote in this journal there was "no form of government that we could perceive," which meant nobody to with or to make treaties or diplomatic agreements with. The intelligence report Woodbury provided to Downes about Sumatra stated that the Acehnese were "without . . . any civilized principles of government." It is likely Downes did not think he could deal with Kuala Batu or other local villages diplomatically, since the Navy Department told him this lack of government precluded "open regular diplomatic relations with the rest of the world." This shows that for operational success, or for naval irregular warfare to affect the strategic level, it sometimes required leadership that understood and placed value on the diplomatic elements of these missions. As Porter's issues in the Caribbean counterpiracy campaign and the next chapter's study of further Southeast Asian operations demonstrate, this was particularly true during peacetime maritime security operations. The historic role of naval officer as diplomat appears critically important.[87]

The platforms and equipment available to Commodore Downes also played a role in the mission to Kuala Batu. The ability to disguise *Potomac* as a merchant was an important part of Downes's operations, but the size of the ship also made this difficult. The landing force at Kuala Batu also benefitted from *Potomac*'s ability to deploy maneuverable fire support. The 6-pound artillery piece and the gunboat, which the crew built out of the ship's launch, were important to the force's reduction of the largest of the forts. *Potomac* could have been used, but as was demonstrated by the destruction of the fort on the far side of the Stone River the following day, a heavy bombardment had a different effect and different accuracy than the smaller guns used during the assault. The mobility of the gunboat under Midshipman Godon's command also provided Lieutenant Shubrick with a maneuver element, which could become defensive as well as offensive during the battle. Naval irregular warfare commonly required mobility and shallow-water capabilities. Through their own ingenuity, the crew of *Potomac* was able to build their own gunboat, but the fact that large warships did not commonly provide these capabilities made them less optimal for irregular operations.

Finally, the partnerships Downes and *Potomac* developed contributed to the mission at Kuala Batu. John Barry and Po Adam, in particular, demonstrate the importance of building relationships, both with local interests but also across the spectrum of maritime industry. Both men provided intelligence, advice, and guidance to Downes and his officers. Barry proved a vital source of sailing directions and knowledge of the local waters, as Secretary Woodbury intended. However, Lieutenant Shubrick also recognized his action ashore, where he served as a guide for the landing force and fought alongside the sailors and marines, reporting, "[F]rom the knowledge of the place possessed by Sailing Master Barry, and his coolness, I derived the utmost advantage."[88]

Barry's rapid appointment as a warrant officer was relatively common during the Age of Sail; the ability to find specialists and rapidly integrate them into the naval force was an important key to success. A better paid and more highly regarded post than that of an enlisted sailor, the sailing master position also allowed the secretary to differentiate between commissioned officers, who were the navy's combat leaders, and officers who offered very particular skills. Because of this, the navy was able to take advantage of the partnership between the department and the merchant community, providing an important source of information and intelligence.[89]

The relationship with Po Adam demonstrates how vital local partners often are for successful maritime security operations and naval irregular warfare. He was central to the original escape by the survivors from *Friendship* and the recapture of the ship. Once the Americans returned, his assistance as translator, husbanding agent, diplomatic intermediary, and intelligence agent provided important service to Downes. While *Potomac* had shipped a Dutch sailor who spoke the local language in Cape Town, Po Adam provided a guide for local cultural and religious issues, and important knowledge of local leadership and relationships between villages. Had Downes had a more diplomatic aim following the engagement at Kuala Batu, Po Adam's intelligence and help would have been even more important. As the next chapter will demonstrate, it was important to maintain continuing partnerships long after the apparent end of specific operations. Many times, maritime security operations did not provide decisive finality and keeping relationships open was important to follow-on operations.

Reports from the coast of Sumatra remained positive for a time after *Potomac* left the waters to proceed east. American trade increased in the region, and both the import and export statistics from Massachusetts pepper traders indicated that it appeared merchants felt more comfortable.[90] The American press, and reports to the Navy Department, remained clear of piratical accounts for several years as the volume of trade from Salem merchants increased. After *Potomac*'s return home in 1834, Reynolds made the claim in his book that "along the whole pepper coast, since the visit of the Potomac, a remarkable change has taken place in the deportment of the natives."[91] Reynolds's defense of Downes's choice to pursue aggressive action notwithstanding, the number of unmolested voyages recorded in G. G. Putnam's records of pepper traders from Salem do support the claim that *Potomac*'s mission had an impact.[92] However, there was not total security or safety. In 1833, pirates attacked the 200-ton barque *Derby*, commanded by Capt. Jonathan Felt, in the early morning hours of 10 June. However, local friends had warned Felt and an extra watch spotted the attackers' boat before they reached the ship. The crew was mustered, a few shots fired, and the boat disappeared back into the night. It was almost seven years before American traders were the victims of another successful, and again deadly, attack by pirates along the northwest coast of Sumatra.[93]

CHAPTER 8

RETURN TO SUMATRA

The East India Squadron, 1838–1839

From 22 August to 26 August 1838 the Salem pepper trader *Eclipse* lay off Tangan Tangan, approximately twelve miles from the larger trading town of Muckie on the island of Sumatra. Anchored in the tropical heat, Capt. Charles Wilkins had not been feeling well and sent his first mate, George Whitmarsh, with four men ashore to weigh the pepper and send it out to the ship from the jungle village. As the men worked, on the afternoon of the 26th the surf rose quickly and became dangerous. Whitmarsh and his party decided to spend the night ashore rather than risk getting back out to the ship. It was unusual but having traded with the village for almost a week, the ship's crew and the locals had developed a sense of comfort and trust. The watches were set on board *Eclipse* and Captain Wilkins retired to his cabin early due to his illness.[1]

That evening a pair of boats came alongside *Eclipse*, with a dozen men in each and several bags of pepper. The second mate recognized the leader as Libbee Ooosoo. He was the brother of Libbee Sumat, a Muckie man whom Wilkins had hired to serve as a *jeretoulah*, or a local agent to help with the pepper negotiation. Sumat was ashore with Whitmarsh. Since the mate knew Ooosoo, he allowed the men on board. He still insisted on locking up their weapons per the captain's standing orders and the Acehnese handed over their knives and krisses. The mate did not want to wake the captain so the Acehnese milled around the deck, waiting. Around ten o'clock that evening Captain Wilkins came up from his cabin for tea. He recognized Libbee Ooosoo as well, and asked him to have tea with him before they weighed the pepper, which the Acehnese had brought.

The crew set up their pepper scales and Ooosoo complained to Wilkins about the lack of trust the mate had shown when he locked up their weapons. Captain Wilkins ordered the men's *krisses* and knives returned, and one of the crew unlocked the chest that held them and returned the weapons. The Acehnese struck. With their knives in hand, they immediately attacked Captain Wilkins, the second mate, and the captain's apprentice, a young man named William

Babbage.[2] Most of the crew escaped overboard and the attackers began to search the ship. They discovered William Reynolds, the ship's cook, whom the captain had confined for insubordination, held in irons below. In exchange for his life and freedom, he offered to show the attackers the location of the money and valuables. The pirates loaded their boats with several cases of opium, roughly $27,000 in specie, and valuables from Captain Wilkins's cabin.[3] Some survivors of the attack swam for shore, but the second mate and a few others climbed back on board the ship after the attackers departed. He gathered the four men who were left, a number of them wounded like him, and lowered a boat. They set off in search of help and farther down the coast came across the French barque *L'Aglee*, out of Nantes. The Frenchmen took them in and treated the wounded. The three men who had swum for the beach immediately went in search of Whitmarsh and their crewmates ashore.[4]

Whitmarsh asked the rajah they were trading with for help. The locals agreed and armed five boats to accompany the Americans back to their ship. The next morning, they rowed out to *Eclipse* and discovered it abandoned. Whitmarsh ordered the men to make the ship ready for sea. As the handful of remaining sailors readied the vessel for sea, the crewmembers who escaped to the French ship returned. They had armed themselves to retake the ship as well, and arrived alongside to discover their shipmates preparing to sail. What was left of the crew prepared *Eclipse* for sea and then set sail for home.[5]

A month later, the frigate *Columbia* and corvette *John Adams* sailed into the harbor at Colombo. The two ships made up the U.S. Navy's newly formed squadron for the East India Station, under the command of Commo. George Read. The squadron was on a regularly scheduled deployment to show the flag and maintain relations on the East India Station, which consisted of the east coast of Africa to the shores of China. The two ships deployed from Norfolk, Virginia, in May 1838 and, after initially following a similar course as *Potomac* at the start of the decade, left Cape Horn bound for the coast of Oman. For several months, the ships made port calls and diplomatic visits, with *Columbia* reaching Muscat and Bombay and *Adams* visiting Zanzibar before meeting *Columbia* again on the Indian coast. Together they visited the Portuguese colony of Goa, on the west coast of India, before heading to Colombo on the island of Ceylon.[6]

Columbia was a 1,726-ton frigate with 50 guns and a crew of 480. The navy laid her keel in 1825, but the ship had not launched until 1836. The navy spent two years fitting her out, and she deployed with Lt. George A. Magruder serving as first lieutenant and Read on board as both captain and squadron commander. Like *Potomac* before her, *Columbia* was a new ship on her maiden voyage on the far side of the world. *John Adams*, on the other hand, had a long and distinguished career when she sailed for the East Indies. A 540-ton frigate, she was armed with

30 guns and manned by a crew of 220. She sailed alongside *Columbia* under the command of Cdr. Thomas W. Wyman.[7]

George Read had a history of combat service in the U.S. Navy nearly as long as *John Adams* did. President Thomas Jefferson signed his appointment as a midshipman in 1804, just missing service in the First Barbary War. However, he rose through the ranks and served as a lieutenant during the War of 1812. Under Capt. Isaac Hull on board USS *Constitution*, Read led gun crews during the fight between *Constitution* and HMS *Guerrière* in 1812, and he was the lieutenant sent on board the British ship to accept the surrender. His next ship was *United States*, where he served under Capt. Stephen Decatur during the frigate's victory over HMS *Macedonian* in 1813. By the end of the war, he had command of his own ship, the brig USS *Chippewa* (16-gun), which he sailed to the Mediterranean with Bainbridge's squadron during the Second Barbary War. He was promoted to captain in 1825 and received the largely honorary title of commodore to take the East India Squadron to sea.[8]

Columbia and *John Adams* began their port call at Colombo with a full schedule of diplomatic activities. Commodore Read and the officers shared formal dinners and socialized with the British officers and diplomats of the British East India Company's settlement. During their visit, the *Colombo Observer* published an account from another British colonial newspaper, the *Penang Gazette*, reporting the attack on *Eclipse*. John Revely, who was the United States consular agent at Prince of Wales Island, sent the story to the editors of the newspaper. It included an account from the captain of the French barque that had rendered assistance to *Eclipse*'s crew, and another from an American captain who was also on the Sumatra coast. The editors in Colombo rhetorically asked, "[P]erhaps Commodore Read may be induced to bend his course, with the *Columbia* and the *John Adams*, in our Roads, to Sumatra, to avenge the death of his countrymen." Read received a nearly identical communication confirming the reports directly from Revely the next day, and the news spread among the crew. Read ordered *Columbia* and *John Adams* to recall the crews from liberty and to prepare to sail the next morning. Protecting American citizens in far seas was one of the key missions of the East India Squadron and the other small American squadrons deployed to stations around the world.[9]

The governor at Colombo held a banquet for the commodore and his wardroom the evening Read confirmed the news from the coast of Sumatra. Not wanting to abandon his diplomatic responsibilities as the senior officer on the East India Station, Read invited the governor and his wife to come on board the ship for breakfast the next morning. He knew his duty required him to investigate the news from Sumatra, but decided he had time for diplomatic nicety as his men finished preparing to get under way. At noon on 1 December, the governor and his wife left the ship after their breakfast, with the crew at their stations and manning the yards. As soon as the party's boat was clear of the ship, Read gave the order to weigh anchor and *Columbia* departed the roads with *John Adams*. Read left behind a letter to Secretary of the Navy James Paulding, reporting the alleged attack on *Eclipse*, stating, "[T]here was nothing to detain us and I sailed on the evening of the 1st of December for the coast of Sumatra."[10]

The passage from Ceylon to the coast of Sumatra took the squadron nearly three weeks. Knowing that the possibility of a naval war in the Far East was slim, and the chances of a squadron or even individual ship engagements were unlikely, Read spent time ensuring his men had regular training with their small arms and in tactics of landing and boarding operations. During the passage to Sumatra practice sessions picked up intensity, and the crew conducted drills with the guns as well as muskets, pikes, and pistols. The officers assigned the crew to divisions for a landing force, just as *Potomac* had done years before. Lt. D. D. Baker, who served in the swamps of Florida in 1837 during the Second Seminole War alongside Lt. Alvin Edson, commanded the squadron's marines and set about using his men to get the sailors into fighting shape. The marines hoisted out targets for a makeshift firing range and the men sharpened the cutlasses, battle axes, and pikes. William Murrell, a member of the crew who published his recollection of the cruise in 1840, wrote of the crew focused on their training "being in full expectation of having a *bit of a brush* [*sic*] with these Malayan desperadoes, on our arrival."[11]

The squadron made landfall on 16 December about fifty miles south of the northern tip of Sumatra and headed southeast along the coast. On 21 December, with *John Adams* in the lead, the squadron approached the village of Annalaboo. The lookouts on board *John Adams* spotted a pair of brigs about five miles from the town, and the two American warships came to anchor nearby. The brigs were under the British flag, and Lieutenant Magruder took a boat from *Columbia* to inquire for intelligence. The traders had just arrived from Penang and had no information beyond the same reports *Columbia* read in Ceylon. One of the merchant captains warned that the locals were known to be treacherous and could not be trusted. When Magruder asked about the harbor at Muckie, the Englishman advised him that the water alongside the peninsula where the town sat was deeper than most of the Sumatran shoreline and the two warships would be able to get in very close for a bombardment, if it came to that. The boat returned to *Columbia* with the limited intelligence.[12]

At three o'clock the next afternoon, the squadron was under way for Kuala Batu. Commodore Read reasoned that, if they were to start anywhere, a pair of

American warships was likely to have the largest visual impact off the shore of the village the U.S. Navy had destroyed a few years earlier. He hoped that, since they had no indications Kuala Batu was involved in the attack on *Eclipse*, they might be able to gather information there. That night the ships dropped anchor within sight of the village.[13] Light and varying winds the next morning kept the ships at anchor. As they waited for a favorable breeze, the lookouts spotted a canoe approaching *Columbia* from the shore. When the small craft reached the side of the ship, a local man climbed on board and announced Po Adam, who was ready to come out to the ship if the commodore desired it, had sent him. The visitor confirmed the reports of the attack on *Eclipse* and claimed one of the pirates who attacked the Americans as well as several thousand dollars of the stolen money were in Kuala Batu. He reported other members of the pirate band were in Muckie and in the village of Soo Soo.[14]

Another canoe was soon spotted paddling toward the ships and in a short time Po Adam himself, the local guide who helped Captain Endicott on board *Friendship* and Commodore Downes on board *Potomac* in 1831 and 1832, boarded *Columbia*. The officers knew of Adam, and of his role in the events on the Sumatran coast at the start of the decade. He greeted them as old friends and the wardroom responded in kind. Adam repeated what his messenger had said, confirming that the rajah at Kuala Batu had two thousand dollars from *Eclipse* and that one of the pirates was living in his village. He also provided detailed information from the attack, including who planned it and which rajahs sanctioned it. He gave the crew details on the identities of the perpetrators and told the commodore the leaders of the attack had organized it in Muckie and the ringleaders were there, with other participants in Soo Soo.[15]

Because there was just one of the attackers in Kuala Batu, and the rajah only seemed to be involved in dividing the spoils following the attack and did not appear to have been involved in its planning and execution, Commodore Read elected to continue with his plan to begin there. Po Adam assured him a party sent to open a dialogue with the rajah would be safe from harm. While there was some suspicion of his advice, because of the repeated warnings of the Englishmen *Columbia* had come across regarding the untrustworthy character of the locals, Read decided to put a boat over the side and send a party ashore led by Lt. James S. Palmer.[16] As evening approached the ship's cutter rowed the three miles to shore with a small number of officers and sailors. They approached the shoreline and met a reception similar to the one *Potomac*'s first boat received in 1832. A large group of well-armed and aggressive-looking men gathered on the beach, brandishing their krisses. Unlike in 1832, however, the Americans were in uniform and openly armed, and Po Adam was with them and continued to reassure them. Palmer was not intimidated and "the boat was driven boldly upon the beach, and the three officers jumped, without hesitation, into the midst of this wild and armed multitude."[17]

Surrounded by the throng on the beach, the Americans pushed forward through the crowd toward the rajah's reconstructed main fortress, with Po Adam leading the way. The party moved through two gates in the stockade, and the rajah's men blocked the majority of the mob that was following them. The Americans found the rajah on an elevated platform awaiting their arrival. Po Adam passed the message from Lieutenant Palmer that the party was there to communicate the commodore's greetings and his desire to talk with the rajah about the recent piratical attack. The rajah suggested the group move into his inner chambers, apparently to avoid the number of villagers who had slipped into the fort to watch the proceedings.[18]

The meeting in the inner chamber was brief, since the landing party wanted to return to the ship before the approaching sunset. Palmer explained that the commodore and the government of the United States wanted to remain friends with the village of Kuala Batu. However, after the attack on *Eclipse* that friendship was in doubt. Palmer insisted the rajah turn over the pirate who was living in the village and return the two thousand dollars that was allegedly in the village. The rajah professed his continued friendship, but he expressed doubts about being able to take the pirate into custody. The man had many friends in the village, and others who feared him. Any of them could warn of an attempt to capture him. The rajah promised he would try and take him that night, while he was asleep. Palmer and his party returned to *Columbia* without incident and reported the meeting to Commodore Read and the rest of the wardroom.[19]

Palmer and his party related that the rajah was courteous and seemed friendly. Po Adam pointed out that he had admitted the attack occurred and even admitted one of the pirates was in the village, which in many ways was unexpected. While it was a significant diplomatic point and suggested a desire to remain on good terms with the United States, none of the Americans believed that the rajah would actually take the man into custody. There was a good chance he would not even attempt it, since it was likely the perpetrators had spread a portion of the two thousand dollars of specie around the village.[20] Commodore Read began planning for the next day with the assumption that the rajah would not deliver. He organized a second party to visit the fort, this time under the command of Commander Wyman, commanding officer of *John Adams*. That night Read wrote out orders for Wyman, which instructed him to repeat the conversation Lieutenant Palmer had had with the rajah, emphasizing the desire of the government of the United States to remain at peace but making it clear the rajah would decide that through his own actions. The orders also instructed Wyman to give a deadline of sunset the following day, 24 December, for the rajah to produce the pirate and money.[21]

Wyman's party went ashore the morning of 24 December. Two officers accompanied him from *John Adams* and three from *Columbia*, including Chaplain Fitch Taylor. Fifty village men met them again on the beach, as the night before, but with their weapons sheathed and a less menacing appearance. The party returned to the rajah's fort and the village leader greeted them with a wave and by shaking hands with Commander Wyman. The Acehnese led the Americans away from the central audience area and into private quarters for the discussion with the rajah.[22] Wyman opened the discussion, with Po Adam and a sailor from the ship who was familiar with the local language serving as dual translators. The rajah claimed he had been unable to take the pirate because he had escaped in the night. However, he had dispatched a group of his men to pursue him and carry messages to other local leaders in order to take him into custody. Wyman delivered the ultimatum Read had set, stating the rajah had until sunset that night to produce the murderer and the money. If he did not, he would be responsible for breaking the peace between his village and the government of the United States. The commander offered him the opportunity to come out to *Columbia* and hear from the commodore himself, but the rajah declined and promised he would send word by nightfall.[23]

The audience complete, the Americans returned to their boats by a meandering route through the village. As the officers observed the forts, a local man approached Chaplain Taylor. The American recognized him as a lower chieftain that had attended the audience with the rajah. He introduced himself as Po Nyah-heit and he assured Taylor he wanted to be friends with the United States, and that his clan had not fought against *Potomac*. To demonstrate his friendship, he offered the Americans a water buffalo as a gift. Taylor offered Po Nyah-heit the opportunity to come out to *Columbia*, however, and offer the gift to the commodore himself. With Commander Wyman's agreement, and repeating of the invitation, the chieftain agreed.[24] Once on board he pledged his friendship to the United States to Read, and told him the rajah not only had the money, but also could have taken the pirate if he wanted. He asked if he could bring his family on board the ship if hostilities were going to begin. Before answering, the commodore asked him what he would do if the ship destroyed the town. The man replied he would return to the village with his men and take control, he would become the new rajah, and "I be friend to America." He even offered his men to fight alongside an American landing party and his fort as a base of operations.[25] The officers consulted with Po Adam about the man. Adam liked the idea and suggested that with a little support from the Americans it would probably work. However, Commodore Read declined Nyah-Heit's offer of a direct alliance. He did offer to take the man's direct family on board, in the case of "any difficulties," but made it clear he made no promises to spare his property or his fort from destruction. The minor chief left the ship.[26]

After noon, Commodore Read and the officers suspected the rajah was not going to respond to the ultimatum Commander Wyman had delivered. The crews kedged *Columbia* and *John Adams* in toward the village, casting out anchors and pulling the ships closer. They put out spring hawsers in order to maneuver the ships' broadsides, aiming them toward the main forts around the town. *Columbia* beat to quarters, the gun crews taking their positions and marines climbing into the rigging for a bombardment.[27] The crew watched from their positions as women and children streamed out of the village, and a pair of heavily laden proas departed from the river. After a few minutes at their positions, the crew was surprised to hear the retreat called. They left their guns in a condition of readiness, and the boatswain piped the sailors and marines to dinner. The crew ate and, as "the officers were nearly finishing their desert [sic]" the commodore again ordered the ship to quarters. Read decided the rajah of Kuala Batu was not going to comply with his demands, and as sunset approached, he ordered his ships to open fire on the village.[28]

Columbia opened with the first shots and *John Adams* followed on that signal. Villagers who had gathered on the beach scattered, and the rajah's westernmost fort returned fire toward *Columbia*. Three shots came from the local fort and, while they were well aimed, all fell short of the ship. The divisions on board both ships continued to fire, clearing the foliage in and around the village, striking the forts, and raising smoke from the bazaar. After half an hour, the commodore ordered a ceasefire.[29] An early evening shower passed through, and the crew observed from the decks of *Columbia* as groups of village men began coming out of their hiding places with muskets and blunderbusses. The officers suspected the local men had laid an ambush for any landing party the ships might send. The forts raised two white flags at Kuala Batu, and the crews could see flags farther down the coast at Soo Soo and Pulo Kio.

The following day was Christmas Day, 1838. The ships spent the day quietly at anchor, some of the men enjoying the holiday and others lamenting the fact the commodore had not ordered a landing party ashore to destroy the town. On 26 December, the officers briefly communicated with some of the Acehnese ashore as white flags continued to fly from a number of forts. Po Nyah-heit and a relative of his, Po Kwala, again offered to fight alongside the Americans if they planned to land.[30]

In the aftermath of the Christmas observances, Commodore Read considered what a landing party sent to destroy the village might achieve and weighed it against the possibility of what appeared to be a planned counterattack. He decided for the moment that the squadron's work at Kuala Batu was complete. He ordered the ships to sail for the village of Muckie, where Po Adam claimed the ringleaders of the attack on *Eclipse* were living and where it had been sanctioned. The ships progressed up the coast slowly, under light winds, and came to anchor off Muckie on 30 December.[31] Muckie was a much larger village than Kuala Batu, referred to in an officer's account as a "city" at one point, and it was situated on a peninsula with steep mountains rising from the landward side. A number of forts stood around the village, the largest on the very end of the peninsula, and cultivated fields extended beyond the buildings at the opposite end of the village and up the beginning of the mountain slopes. Lush coconut groves in and around the village, fields and pepper vines, and rising mountains all contributed to a tropical beauty.[32]

Read sent Commander Wyman ashore to open a dialogue with the rajah of Muckie. Over a hundred men gathered on the beach, just as the Americans had first encountered at Kuala Batu. Both of *Columbia*'s cutters, filled with armed reinforcements, accompanied Wyman and his party and waited just off the beach as the negotiators landed. Wyman explained to the village leaders who met him that the American ships were investigating the attack on *Eclipse*, and as he had at Kuala Batu he demanded local leaders turn over those who took part in the attack and any of the money which was in the village. He invited them to come out to the ship to discuss the matter with the commodore. The rajah and his party refused the offer that afternoon but suggested they would come out the next morning.[33]

The morning of 31 December Lieutenant Turner, *Columbia*'s senior lieutenant, went ashore with the boat in order to bring the delegation from the village out to the ship. However, the villagers again declined the invitation. They claimed they feared for their own safety, but the rajah also insisted he had nothing to do with the piracy. He said he was not responsible for what people who lived in his village may, or may not, have done to foreigners. Lieutenant Turner returned to the ship and passed the message to Read, as well as his opinion from the meeting that there would be nothing gained from continuing to negotiate with the rajah or his people. Read concluded the same and later wrote: "[S]atisfied that they did not mean to comply with my demands for the persons concerned in the piracy, I directed all preparations be made." The crew set to work to haul the ships in closer to shore and prepare for bombardment.[34] Early the next morning Wyman began moving *John Adams* closer to shore and *Columbia* followed. With the information on water depth that the Americans had received from the British brigs when they approached Sumatra, they confidently towed and warped the ships into position approximately two hundred yards from shore. The crews put out spring lines and pulled their starboard batteries to bear on the village. Read determined that the leaders of the village had no intent of cooperating. As the ships maneuvered toward their firing positions, the crew observed a steady stream of the village's inhabitants carrying off their valuables and disappearing into the jungle-covered mountains. Both ships were in position for bombardment by 1000.[35]

Read beat his ship to quarters and at 1100 he ordered his squadron to open fire. *Columbia* began firing with selected guns from her main deck and spar deck, and *John Adams* took up the signal and joined. Read reported that once the cannonade began he no longer observed the villagers trying to escape, and he continued the bombardment in order to prevent the preparation of any kind of defense from the village. *Columbia* focused her fire on the forts, and both ships rotated their round shot with canister and grape, which shredded the coconut groves and thinly constructed buildings around the village bazaar.[36] As noon approached, Read ordered a cease-fire. After an hour of bombardment, silence stretched out over the wreckage of Muckie. The commodore ordered his landing party called away. Six boats carrying three divisions of men from *Columbia* under the command of Lieutenant Magruder, and four from *John Adams* carrying two divisions, embarked their men on the seaward sides of the ships. Wyman boarded one of the boats from *John Adams* as the expedition's overall commander, and Lieutenant Baker boarded one of the *Columbia*'s boats in command of the squadron's marines. Commodore Read addressed the men from *Columbia*, looking down on the boats after the force had embarked: "You have been desirous to have an opportunity to land, on an expedition like the one that is now offered to you. I have the fullest confidence in your success. Burn and destroy the town, and put to death all men who you may find bearing arms, and by no means injure the unarmed or yielding. Gentlemen, I wish you success and shall expect your return in one hour and a half."[37] In total 370 men were embarked, and the boats set out for shore as the remaining gun crews on board *Columbia* sent occasional shot toward the beach to protect the landing force.[38]

The boats landed at the harbor's edge, about 150 yards from the village itself. The marines landed first, and Lieutenant Baker immediately led them up the small sand ridge from the beach and took up defensive positions facing the town. The divisions cleared their boats and began to form up on the beach. As the landing force moved ashore, the petty officers in charge of the boats rowed them back off the beach, anchoring a short distance away. By 1230 the force was formed, Wyman delivered orders to his officers, and they were ready to march for the town. A pair of red flags flapped on two poles at the edge of the sand ridge, before the large open area between the beach and village. The purser from *John Adams* approached Commander Wyman and asked for permission to take the flags before the force marched. Wyman agreed, but sent a guard with him. To the surprise of the Americans, who expected sniper fire on the attempt to take the flags at a minimum, the purser took the flags without a shot fired, retiring back to the beach. Instead, on his return, the officer reported he had seen through the village to the other side of the buildings where a stream of villagers continued heading toward the mountains. It was Wyman's first indication that the operation, instead of facing an ambush, might be less violent than feared.[39]

Similar to Downes's plans at Kuala Batu, Wyman split off his divisions to assault the village's forts. However, unlike six years prior at Kuala Batu, they were largely empty and provided no resistance. The ambushes that the marines feared as they entered the stockades did not materialize, and as the forts fell to the landing force the guns were quickly and easily spiked.[40] A division was broken off with torches to begin firing the bazaar, which they also discovered abandoned and cleared of all valuables. As the main force moved to the final fort and a redoubt, which protected the rajah's home, one of the officers pointed out a building that appeared to be the village mosque. Wyman ordered the structure spared, and the torch-men moved along toward another section of buildings.[41] Inside the elaborately carved gate of the main fort and rajah's home the sailors and marines found three abandoned buildings. The lieutenant that took a detachment up and into the redoubt reported discovering five guns, which were unmanned. The party removed the guns from their mounts and spiked them, and the structure was set ablaze like the others.[42]

With all five of Muckie's forts taken and twenty-two cannon spiked and dismounted, the divisions began to reassemble. The detachment of men in the boats just offshore spotted a large amount of movement at the edge of the village. The officers were well aware that during both British and Dutch operations on Sumatra, defenders had hidden and then counterattacked landing parties as their operations drew to a close. The attacks resulted in catastrophic losses for the Europeans. Sailors and marines marched for the edge of town to confront the enemy. Spotting movement in the distance, they took positions to engage the counterstrike, presenting what one memoir recorded as "really an imposing sight, [which] appeared unexpectedly martial and formidable."[43] As the Americans drew closer to the movement what they discovered, instead of a defensive counterattack, was a herd of water buffalo, which stampeded away after spotting the military formation.[44]

By 1400, Muckie was in the possession of the Americans. Wyman reported to Read that the town was consumed in flame. As "Yankee Doodle" and "Hail Columbia" sounded on the bugles, the divisions reformed on the beach. A last batch of local watercraft and a proa under construction were set afire as the Americans embarked their boats and began returning to *Columbia* and *John Adams*. Without a single casualty, the Americans returned to the ships and by half past 2 o'clock, the operation was complete.[45] Wyman wrote, "nothing remained visible to the eye but the ashes covering the smoking ruins upon the site on which the town of Muckie and the forts once stood."[46]

Commodore Read wrote in his report to Secretary Paulding that, while the landing force had encountered no opposition, he was confident they would have handled anything that might have come their way. Particularly, he stated, "much was expected of the marines on this occasion, and much no doubt would have been done, had further proof of their skill and discipline been required." Read ordered the ships warped back into deeper, safer, water and "spliced the main brace for the men," delivering an extra ration of rum, as evening set in. In the morning, he ordered the squadron to weigh anchor and sail down the coast, past Kuala Batu, to the anchorage off Soo Soo.[47]

The morning after arriving in Soo Soo, Read and Wyman sent parties ashore to begin watering, just as Downes had with *Potomac*. The men went armed with pistols and cutlasses, and the marines again set up perimeter positions on the bluff overlooking the small river. Just as during *Potomac*'s visit, several armed local men arrived to monitor the watering operation, but they remained outside of the marines' line. The American sailors ran watering parties all day, while Read and his officers opened negotiations with the rajahs of Soo Soo.[48]

Po Adam and others confirmed there was at least one pirate residing in Soo Soo, maybe as many as four. A part of the money taken from *Eclipse* was likely in the village as well.[49] Read arrived off their beach with the intention of inflicting "a moderate castigation upon the inhabitants for permitting several of the pirates to reside among them." The Americans demanded that the locals turn over the pirates, but they were well aware it was unlikely to happen because neither Kuala Batu nor Muckie's leaders had complied. The watering parties reported they were having no trouble ashore, despite armed onlookers from the village. Read observed "the civil and quiet manner in which they conducted themselves" and decided a more diplomatic route might be better.[50] It was obvious the Americans were capable of levelling the village. After landing with Po Adam to tour the settlement, Read observed the forts in Soo Soo were in the worst condition of any they had seen and the rajahs surely knew of the weakness of their defenses. The local leaders assured him they wanted to comply with his instructions to turn over the pirates, but they did not have the power to do so. Read decided their response was likely accurate, and violence would not have the impact the Americans desired.[51]

The negotiations with the "priests of Soo-Soo," as Chaplain Taylor referred to them, began by using them as intermediaries for a dialogue with Kuala Batu. When *Columbia* and *John Adams* sailed down the coast from Muckie they spotted numerous white flags, and the officers believed the whole section of the coast had heard the news of Muckie's destruction and now understood the power wielded by the Americans. Unlike Commodore Downes, Read was not satisfied with conducting a military operation and then sailing away, assuming the locals had learned their lesson. He intended to pursue something more lasting, including diplomatic negotiations that could lead to treaties.[52] The rajah of Kuala Batu did not want the Americans to return to his village, after hearing about the razing of Muckie. He was also a local merchant who understood the economics involved in the situation. Muckie, the largest of the local trading villages, lay in ruins and would not be open for business until they were able to rebuild. That left Kuala Batu in a position to improve their standing in the pepper trade, so long as the rajah could keep the Americans from levelling his village as well. The potential for economic gain, or loss, for the village came into stark relief when two European traders arrived on the coast and Read had them met by his boats. The Americans turned the merchant ships away, insisting they leave until the naval officers had concluded their negotiations. The message to the local leaders was clear.[53]

The rajah of Kuala Batu sued for peace between his village and the United States. While he did not acknowledge receiving any of the pirated money, he did concede two thousand dollars of it came to his village. He was unable to place the pirate into American custody, but he offered to repay the two thousand dollars in order to demonstrate his good intentions. Since he was unable to gather that amount of money on short order, he promised to pay the sum in one year's time to any American vessel designated to receive it. His village stood to make far more than that sum over the following year, with Muckie removed from the pepper trade, and the Americans realized the commercial wisdom of his offer. Read wrote to Secretary Paulding that since, at the time, he had neither pirates in custody nor money in hand and laying waste to the rest of the coast appeared possible but unproductive, he accepted the offer.[54]

Read had a document drawn up that described the events that had taken place, including the attack on *Eclipse* and bombardment of Kuala Batu. It then enumerated the rajah's offer to pay the sum in one year, and Read's agreement that it satisfied the American government. Both the rajah and the commodore signed the document.[55] Read reported to Paulding that word from Soo Soo was that Kuala Batu considered it the same as a peace treaty between their village and the United States.[56]

Once the negotiations were complete with the leading rajah of Kuala Batu, and other local leaders heard about the document signing, Read was approached by numerous requests for similar documents. After turning away the European pepper traders, it was clear Read was not just thinking militarily; as a good naval commander he was also thinking about the economic impact of his maritime security capabilities and the diplomatic implications.[57] In particular, he believed he needed to negotiate with Soo Soo to establish an understanding before his squadron left the coast. Po Adam lived in Soo Soo, and as Taylor related, the Americans felt "we owe, as it is supposed, something to Po Adam."[58] The unnamed officer and memoirist of *Around the World, a Narrative of a Voyage in the East India Squadron* agreed, writing that the wardroom believed "the kind and useful services of Po Adam certainly deserve a more substantial return than any mere pleasures of benevolence which he may inwardly enjoy."[59]

With the continued freedom of the pirates in or near the village, however, Read felt his squadron could not just leave the scene and allow Po Adam to claim responsibility for the Americans' mercy. Po Adam reported from the village that, after the bombardment of Kuala Batu and the departure of the ships for Muckie, the village thought they were safe. However, the return of *Columbia* and *John Adams* resulted in great fear, and "the women cry, and the men too when the big ships come again."[60] On 7 January, Read, Wyman, and a party of officers including Chaplain Taylor, took the commodore's gig ashore to Po Adam's compound to discuss a treaty with the rajahs of Soo Soo.[61] Upon the Americans' arrival, Po Adam's servants produced coconuts and split them to prepare drinks for the group. The two principal rajahs of Soo Soo arrived and joined them in the customary refreshments. Read produced the document, which the Americans had prepared on board *Columbia*, for review by the rajahs. It opened, stating: "We the Rajahs of Soo Soo, for ourselves and the inhabitants of Soo Soo on the west coast of Sumatra, sensibly affected by the clemency practiced toward us, on the late visit of the frigate Columbia and the John Adams, do hereby pledge ourselves to suffer no American vessel to be molested hereafter, and by all means in our power to prevent all wicked designs for annoying or in any way injuring them."[62] The treaty went on the pledge that the local villagers would warn any Americans of piratical attacks if they learned of them prior to the event, and they would come to aid in the defense of American vessels that were attacked. If an attack were to occur, the rajahs promised to capture the perpetrators and hold them for the first American warship to arrive on the coast.[63] The rajahs signed the document and in broken English the leaders assured the Americans "Soo-Soo safe now, we no fight now—we friends" and the men shook hands.[64]

After the squadron's boat crews turned away the two foreign traders, an American merchant vessel named *Sumatra* joined *Columbia* and *John Adams* at anchor and her master began inquiring about a cargo of pepper. Capt. Peter Silver of *Sumatra* had traded on the coast a number of times and told Read he had never seen such fear from the local villagers. After Read signed the agreement with the leaders of Soo Soo, Silver brought an offer to negotiate another treaty from three minor rajahs in the surrounding area. Silver warned the commodore about one of them, named Po Quallah, from the small settlement between Soo

Soo and Kuala Batu called Pulo Kio. The village had been the home of Po Adam at the time of *Potomac*'s operations on the coast. However, the other local leaders had displaced Adam following his cooperation with the Americans in 1832. Silver characterized the new rajah that oversaw the settlement as "the most designing and dangerous chief."[65]

Read sent Lt. Alexander Pennock and Chaplain Taylor ashore with Silver to negotiate with Po Quallah. Arriving at the settlement, made up of a small bazar of about fifteen or twenty shops, a pepper storehouse, and a single fort, the group met Po Nyah-heit, who had come on board *Columbia* prior to Kuala Batu's bombardment. He was one of the minor rajahs who pledged loyalty to Po Quallah rather than the rajah of Kuala Batu. After a period of tedious waiting, likely meant to demonstrate his importance, Po Quallah arrived at the storehouse to begin discussions.[66]

Taylor commented in his memoir: "[H]is person was rather small, his deportment more gentlemanly than any other Rajah's I have met with." Taylor assumed the role of lead negotiator. After the rajah had taken his seat at the head of the storehouse's veranda, servants passed the coconuts around for everyone to drink. Through the sailor who had been serving as an interpreter during the squadron's diplomatic negotiations, Taylor explained to Po Quallah, "[T]he Americans desire to be on friendly terms with the Rajah and their people"; however, it had twice now been necessary for American warships to come to the coast and inflict punishment for piracy. Taylor warned that in due time more warships would visit the villages, and if more acts of piracy occurred the locals could be sure there would be more punishment. He then laid out the same terms that made up the document with the rajahs of Soo Soo, and the same pledges to help American vessels.[67]

Taylor had the exact language in the prepared document translated to him, and Po Quallah agreed to the terms. He produced his stamp and signed and stamped the document. The agreement bound Pulo Kio to assist all American vessels in trouble along the coast, including giving them any intelligence of planned or proposed pirate attacks and assisting them with force, if necessary, to repel attacks already in motion. Taylor and Lieutenant Pennock, and three of the lesser rajahs, including Po Nyah-heit, signed two copies. Four members of the crew signed as witnesses.[68] With the document signed, Po Quallah agreed to visit *Columbia* and he and his men boarded a boat to the ship, and Taylor introduced them to the commodore and Commander Wyman. Read signed the copies of the document and, after a tour of the ship, the party returned ashore with "the Rajah expressing his high gratification."[69] In the classic result of gunboat diplomacy, Captain Silver and the crew of *Sumatra* contracted for a cargo of pepper with Po Quallah on terms that seemed to benefit everyone present—though clearly not the European merchantmen driven off by American boat crews.[70]

The U.S. East India Squadron departed the coast of Sumatra on 14 January 1839, northbound for the Strait of Malacca, en route to visit Singapore. They spent another year and a half deployed, rounding Cape Horn on a circumnavigation, and returning to Boston harbor on 13 June 1840. At the end of the deployment *Columbia*'s logs showed nearly 55,000 nautical miles travelled in defense of American interests in the Indian and Pacific Oceans.[71] American pepper merchants would dominate the trade on the Sumatran coast through the 1850s. There would not be another major pirate attack on an American vessel through the Civil War and the end of American domination of the pepper trade.[72]

Commodores John Downes and George Read faced very similar situations in 1832 and 1838; the geography, the details of the attacks on the merchant ships, and the populations involved were all virtually identical. Yet the two men demonstrated distinct mind-sets as they approached the coast. While the tactical results of their decision making on the shores of Sumatra appear similar, resulting in bombardments and combat landings, they came to those results from very different approaches. At the end of their time on the Sumatran coast, despite the same physical results in terms of villages burned and populations chastised, the two expeditions had quite different diplomatic and strategic results.

It may be that the most significant difference between the two events, and the one that had the most important impact on the outlook of the commanders, was that in the orders they received from the secretary of the navy. As squadron commodores, both men initially received similar general orders from their respective secretaries. During the antebellum period, the orders provided to squadron commanders tended to be as much diplomatic as they were military: detailing the state of American relations in the region where the squadron's ships would cruise, as well as military considerations of other naval forces in those waters. Downes received his orders from Secretary Woodbury with the caveat, "[C]ases may arise which it is impossible to foresee, and to meet which definite instructions cannot be given; should such occur, out of the ordinary way, you must be left to the exercise of sound discretion."[73] Read received similar language from Secretary Mahlon Dickerson.[74]

However, following the arrival of *Friendship* back in Salem and Secretary Woodbury's correspondence with the owners, he issued a second set of orders to Downes specifically addressing his instructions for the coast of Sumatra. The second set of orders had three elements. First, upon arrival on the coast of Sumatra, Downes was to gather as much information as he could about the events from Americans trading in those waters. Second, he was to demand that the rajah of Kuala Batu provide restitution and punish those responsible. Third, if the rajah refused, Downes was to take custody of the responsible parties himself and "destroy the boats and vessels of any kind engaged in the piracy, and the forts and dwellings near the scene of aggression, used for shelter and defense; and to give public information to the population there collected, that if full restitution is not speedily made, and forbearance exercised hereafter from the like piracies and murders upon American citizens, other ships-of-war will soon be dispatched thither, to inflict a more ample punishment."[75] Downes was to return the offenders to the United States for trial by whatever method he found convenient and proceed to assume his duties in command of the Pacific Station's ships.

Of the three parts of the orders, Downes focused on the third element. During the trip to Sumatra, at Rio and again at Table Bay, merchants warned Downes and foreign officers there would be no way to trust the local population at Kuala Batu. As he neared the coast, his plan to land and surprise the village, prior to entering into any kind of discussion with the local leadership, appears reasonable if the focus was the second and third elements of the orders, rather than the first two. Downes's explanations to Woodbury, following the completion of the operation at Kuala Batu, make it clear that the execution of the third element of the orders took up a majority of his attention. It was the portion that placed his men and his ship at physical risk, and thus drove how he approached the rest of the instructions.[76]

Almost seven years later, as Read's ships approached the same coast, he viewed his task in a much different manner. His orders were not as clear as Downes's because he was acting on his own in response to a mid-cruise crisis. Yet he was clearly aware of what happened to Downes and focused his approach in the opposite manner. As detailed in the previous chapter, when the news of *Potomac*'s landing reached the United States it became a political firestorm, with outraged newspapers and a congressional investigation called. While in public President Jackson and Woodbury backed up their commodore, in private and within the Department of the Navy their displeasure was obvious and quite strong.[77] Read was well aware that Downes never returned to sea after his cruise on board *Potomac*. Instead, as he approached Kuala Batu in late 1838, Read looked specifically to gather intelligence and open diplomatic dialogues with local leaders. *Columbia* and *John Adams* fell in with a French vessel and stopped to ask for news or intelligence, though they knew nothing more than the Americans did. After making landfall, Read stopped to consult with a pair of British merchant ships who might have had more information, but did not. He also welcomed Po Adam on board and listened to his report, and clearly looked to talk with each group of local rajahs as he moved along the coast. Of course, these negotiations failed, and the result was bombardment and landing in punitive expeditions that look quite similar to the results of Downes's mission as well.[78]

But it was not only prior to the beginning of combat operations that Downes and Read differed in their approaches, but afterward as well. Downes was content with having chastised the population, which he believed condoned the piracy. With his flag of truce flying at Soo Soo, he spent more time giving tours of *Potomac*'s gun deck, in order to encourage awe at American combat power, than he did engaging in more substantive diplomatic discussions. On the other hand, Read and his officers worked the human elements and interests: balancing their friendship with Po Adam and their desire not just to chastise but to genuinely stop piratical attacks on Americans trading with the Sumatran coast. While Read was well aware that the documents he was signing with local rajahs were not treaties, and were something that today might be termed "letters of understanding" at best, he was more than happy to allow the rajahs to describe them in any way that worked well for American interests.[79]

From a tactical standpoint, *Potomac*'s landing was a success. The plan was well conceived and executed by aggressive junior officers. Once the local defenders spotted the landing force and opened fire, the Americans were able to carry the day with minimal casualties and relative speed. Despite the tactical success, however, the mission was less effective at a higher level. Secretary Woodbury saw it as a debacle and failure. By the time Downes and *Potomac* returned to the United States two years later, however, the political leadership and the press had largely forgotten the incident. There was no court-martial for disregarding orders, as Downes's friend and mentor David Porter had experienced. There was not even a board of inquiry. However, attacks on Americans did not stop either. Read's approach appears to have had much more success in deterring attacks on American interests in the region.[80] When Capt. Lawrence Kearny touched at Kuala Batu in October 1841, he discovered a coast that remained friendly toward Americans and continued to be cognizant of its agreements with the U.S. government. There had been no further attacks to report, and Kearny wrote to then Secretary Abel Upshur that he did not anticipate any further threat from the rajahs.[81]

Labelling Read's approach as purely diplomatic is inconsistent with the violence experienced by the population of Kuala Batu and Muckie (though clearly accurate at Soo Soo). However, his ability to consider the balance between violent action, which he believed was required to back his negotiation, and diplomatic engagement demonstrated a different kind of leadership than Downes exhibited. From Edward Preble's constant diplomatic work with the Kingdom of the Two Sicilies and elements of the Royal Navy during the First Barbary War, to David Porter's experience at Fajardo during the West Indies counterpiracy campaign, the understanding (or lack thereof) of the importance of naval diplomacy frequently played a central role in successful maritime security or irregular operations. In his work on naval strategy, Alfred Thayer Mahan described what has been termed a "trident" of naval strategy: the balance between military, diplomatic, and economic elements of power. While Mahan was considering the national-level strategy and creating the idea of grand strategy, comparing Downes and Read indicates that an understanding of that trident is also central to success at the micro-level and in the naval irregular warfare operations of our naval history.[82]

The operations of the East India Squadron in 1838 and 1839 also included aggressive and tactically proficient junior officers. They smoothly executed the bombardments. All involved lauded the landing at Muckie for its speed and efficiency. As in the episodes studied in previous chapters, the marine detachment played an important role in the East India Squadron, not only in operational elements of the mission but also in training that ensured the sailors involved were prepared. As in examples from the West Indies counterpiracy campaign, the operations at Sumatra included junior officers who were directly involved in the execution of the diplomatic elements of their commander's plan.

The patterns in American naval irregular warfare continued into the transition years between the Age of Sail and the steam era, and it continued to play an important role in the operational history of the U.S. Navy. As the nineteenth century continued, peacetime maritime security operations remained important missions for the navy as it deployed globally in defense of American interests. These operations continued the navy's involvement in the nation's military, diplomatic, and economic interests. American sailors and marines repeatedly engaged pirates off China, sometimes in cooperation with their British counterparts, as they had in the West Indies.[83] Following the first encounter and irregular operations against the inhabitants of Fiji during the United States Exploring Expedition led by Lt. Charles Wilkes, American naval forces conducted raiding operations repeatedly in the South Pacific islands during the 1850s in response to attacks on traders and civilians.[84] Irregular warfare also remained an important operational element of wartime naval missions. During the Mexican-American War, naval irregular warfare operations were central to the campaign on the west coast, not only at Upper California but also in the understudied operations of Baja and Lower California.[85] During the American Civil War, irregular operations became a staple of both sides in the conflict, much as partisan and cavalry operations remained important ashore.

CONCLUSION

The Regularity of the Irregular—Raiding and Irregular Warfare in the Age of Sail and Beyond

A century ago, Theodore Roosevelt looked to build a global navy by focusing on a blue-water capability, which, in many ways, the United States lacked. In doing so, and with appropriation of John Paul Jones's heritage and redefinition as the father of a blue-water navy, he set the nation on a path to becoming the maritime hegemon it grew into in the latter half of the twentieth century. However, the narrative that was constructed left behind important elements of sea power, the irregular elements, which did not conform to the ideals of oceanic battle and the navy's self-image. These elements have played central roles in the interaction between global and regional powers across centuries of history and are illustrated in the U.S. Navy and Marine Corps' early operational history. Reassessing the history of John Paul Jones, as well as his successors like David Porter, Francis Gregory, and George Read, redeems their credentials and experience as irregular operators, and introduces today's naval policy makers and leaders to the balance of conventional and irregular missions, skills, and duties that are essential in a global navy.

From the American Revolution to the dawn of the Steam Age, maritime raiding and naval irregular warfare were common and important operations conducted by the United States naval forces. In wartime, irregular warfare and raiding missions achieved specific and targeted military objectives and contributed to the larger operational and strategic aims of commanders who employed them. In peacetime, irregular operations filled the gray area between war and peace when limited combat operations were required to advance national objectives and protect American interests. While the dominant historical narratives of the era have focused on frigate duels in the nation's wars, the how, when, and where of building ships of the line, and the diplomatic and commercial negotiations during peace, naval irregular warfare made up a vital and all-but-ignored element of the U.S. Navy's operational history. While irregular in tactics and objectives, when compared with conventional sea battles, these important naval operations were anything but irregular in their frequency and contribution to American sea power.

The scholarly study of American naval history, now including the expanding examination of the administration, construction, and manpower of the blue-water force, has left large periods and numerous episodes of operational naval history unexamined. A navy's roles in littoral operations and in peacetime remain poorly studied and underanalyzed parts of the maritime past.[1] This is particularly true of the operational history of maritime raiding and irregular warfare, since there has been some effort to begin the work of uncovering naval roles in diplomacy and commerce. Any effort to study irregular operations in peacetime also requires the study of similar operations during war. Because our histories of the U.S. Navy in the Age of Sail document neither very well, this study returned to the archival and contemporary records to establish what actually happened.

The episodes and examples of maritime raiding and naval irregular warfare in the previous chapters suggest not only the intellectual sensibility of guerre de razzia as a concept, but also offer historical and practical credibility to the concept. Comparative analysis across these episodes offers important themes and parallels. Not only does the study of naval irregular operations provide much-needed expansion of the historical record, but it can also help historians and practitioners better understand the elements that contribute to success and failure in these kinds of missions. John Paul Jones, as the father of naval irregular warfare, demonstrated elements of the kinds of leadership required, the connection between irregular and conventional naval operations, and the importance of partnerships or local knowledge or participation in successful missions. These themes carry throughout the cases that progress through the period, which also introduce the role of technology, weapons, and platforms, and deepen the examination of the people and partnerships involved.

The people involved played an obviously critical role in the naval irregular warfare of the Age of Sail. At the founding of the U.S. Navy and Marine Corps, as the Quasi-War developed, Secretary Benjamin Stoddert wrote that he needed senior officers who balanced their bravery with the possession of "good sense," and he needed junior commanders who at the very least exhibited bravery and zeal.[2] Over the course of the Age of Sail, it becomes apparent that "good sense," in relation to the balanced use of irregular warfare, came from experience. The backgrounds of the senior officers involved in successful irregular operations generally included prior leadership of maritime raids as junior officers. This tended to develop commodores and captains who understood and could effectively employ irregular operations once in senior positions. While Commodores Talbot, Preble, and Porter all participated in irregular missions as junior officers and then successfully dispatched them as seniors, John Rodgers remained suspicious of them and focused on the conventional operations, which dominated his own time as a junior officer. However, the experience of John Downes indicates experience with, and belief in, the effectiveness of irregular operations can be taken too far. Success requires the ability to balance between conventional and irregular operations, as John Paul Jones did, but also an understanding of diplomatic relationships as George Read demonstrated.

As a corollary to the operational "good sense" of senior officers, the kinds of orders offered by civilian leadership and the rules of engagement they promulgated raise issues in multiple examples from these chapters. Officers who understood the intentions of their civilian leaders were able to balance the operational aggression necessary for successful naval irregular warfare with the risk of things going poorly. Both Talbot and Porter launched raids into neutral Spanish harbors, but Talbot knew Secretary Stoddert was looking for action and was willing to look the other way, while Porter had already spoiled his relationship with Secretary Southard. Senior officers rarely received orders that were explicit in their rules of engagement when it came to irregular warfare. This required careful risk mitigation and operational planning. Aggressive junior officers, demonstrating the zeal and bravery desired by Stoddert, rose through the ranks to mentor and teach following generations. This was not doctrinal or organizational learning, but instead tacit learning based on ad hoc mentorship, and it resulted in mixed outcomes.

For example, Porter served with Talbot and Preble, learning from the examples they provided, and later Downes and Gregory served with him, observing his example. But this learning was implicit, rather than explicit. The naval officer corps of the day had no standardized or formalized training regimes, and instead learned their profession through the mentorship of senior officers, or observation and experience. As a result, the outcomes varied and there is little documentary evidence to demonstrate learning or training definitively. Layered on top of this informal arrangement, operations on the blue water continued to dominate the "self-image" of the navy, and prize money from captures and victory remained critical motivators. This likely resulted in the development of irregular operational skills and knowledge as an unconscious outcome, occurring through that mentorship and personal experience rather than through formalized training or doctrine.[3]

The ships, weapons, and hardware used within examples of naval irregular warfare in the Age of Sail were important to success and failure in missions. Locating the examples studied in this book explicitly within the Age of Sail helped identify deviations in technology or design more easily. Early innovations with steam power and undersea warfare demonstrated how irregular operations offered an opportunity for early adoption of new and disruptive naval technology.

The missions were small and did not involve fleetwide adoption of new weapons or systems, so it was possible to execute them with less expense and less exposure to organizational risk. This, in turn, resulted in less organizational or cultural opposition. The purchase and deployment of *Sea Gull* to the West Indies gave the U.S. Navy its first experience with steam power in combat operations, where they learned lessons about the challenges of logistics with spare parts and fuel. These would become important elements of the wider adoption of the technology in succeeding decades. The experience with torpedoes and Fulton's introduction of undersea warfare demonstrated that mechanically proven technology was not automatically effective, either tactically or operationally. New technologies required operators or users practiced in and knowledgeable of not only the system itself, but also how it fit into naval operations more widely. Lack of experience in planning and laying torpedo attacks repeatedly resulted in failures like Mix's attempt on *Plantagenet* and the loss of the Berrian torpedo boat.

New technology, however, was not the only element of weapons and platforms illuminated by the history. Throughout nearly every one of the examples, the presence of small craft or minor combatants played a central role. From a repeated need for more shallow-draft cruisers in the Quasi-War, to Preble's efforts to lease gunboats from the Kingdom of the Two Sicilies for operations in Tripoli harbor, to Porter's Mosquito Fleet, and *Potomac*'s construction of their makeshift gunboat *Polly Hopkins*, the need of small combatants for naval irregular warfare is a constant theme throughout the period. Small ships do not normally fit with the imagery of the large blue-water fleet. When navies create fleets based on their aspirational pursuit of glorious and decisive blue-water engagements, small ships fall aside as financial limitations make them appear superfluous. This combined with the political debates over the size, funding, and roles of the U.S. Navy and Marine Corps during the era resulted in a lack of formal procurement policy that would support the small vessels needed for irregular success. The history of irregular warfare in the Age of Sail indicates that, regardless of whether examining wartime or peacetime missions, small ships were a central part of what made up a naval force capable of executing the entirety of its maritime responsibilities.

Finally, relationships between the naval force conducting operations and outside entities played a repeatedly important role in the history. The importance of partnerships fell into three categories. First, partners provided direct intelligence and information on geography, hydrography, and diplo-cultural aspects of naval irregular operations. *Ranger*'s Irish fishermen, Salvatore Catalano's central role in the raid on *Philadelphia*, the contributions of the Matanzas merchants, Po Adam in the repeated operations at Sumatra, all reveal the importance of local expertise obtained through partnerships with non-naval individuals or organizations.

Second, nation-state partners can be important to indirect and logistical support of irregular forces. French basing and support to Paul Jones and *Ranger* in the months prior to the formal alliance, the quiet British assistance provided to Preble in the Mediterranean, and intermittent cooperation from Spanish authorities in the West Indies, all demonstrate that cooperation is necessary beyond formal alliances. Third, combined combat operations repeatedly played roles in naval irregular warfare of the Age of Sail. From the gunboats and crews loaned to Preble by the Kingdom of the Two Sicilies for the attacks on Tripoli in 1804, to combined American and British operations in the West Indies against pirates, conducting direct combat action with partners improved operational effectiveness and provided more militarily effective forces for irregular campaigns. The examples from the War of 1812 demonstrate that the partnership characteristic of joint operations between naval and military forces also proved to be an element of effective operations. The United States Navy and Marine Corps during the Age of Sail was a relatively small force despite its global reach. How, or whether, they used the support of friendly navies, local inhabitants, expatriate or overseas American merchants, and other partners, had a critical role in the operational history of naval irregular warfare.

The examination of the episodes in this book reveals that there is much continuity in naval irregular warfare during the Age of Sail. It involved a different kind of leadership than ships or squadrons maneuvering to position their broadsides, one that necessitated senior officers with diplomatic skill and understanding, and junior officers who were aggressive but could balance that with an understanding of risk in their own commands. It involved different kinds of ships, valued smaller combatants, and commonly required an openness to new technology on a different scale than discussions of fleet design and construction portray. And it involved an ability to integrate and coordinate efforts with partners and allies beyond sharing signal books and coordinating squadron maneuvers; partnerships that acknowledged interests that might have been in parallel, though they were not always the same.

In some of the examples examined in the previous chapters, the naval irregular warfare operations were conducted simply as one more tactical method employed to strike at the enemy. While these cases sometimes had operational and strategic implications, as with the cutting-out mission at Black Rock in the War of 1812, they were not necessarily conducted as strategy per se. However, some of the examples demonstrate that maritime raiding was sometimes used in an explicitly strategic way by the early American naval forces. John Paul Jones and Robert Morris's correspondence over the potential raiding missions to the coast of Africa and West Florida discussed the raiding as a new strategic approach to combating the British. In the West Indies counterpiracy campaign of the 1820s, the shift toward smaller, lighter vessels in order to facilitate raids inshore and irregular operations likewise had strategic implications. Just as commerce raiding could be used as one more tactical effort against the enemy or as an overarching strategy, the same was the case with maritime raiding.

In the first decades of the twenty-first century, Western militaries have confronted challenges of counterinsurgency, hybrid conflict, and irregular warfare throughout the world. In military history, none of these concepts is new. There has been much research and discussion, both practical and academic, on the subjects. A vast majority of recent scholarship and writing on these forms of conflict has focused on today's operations ashore, particularly in Iraq and Afghanistan and in the ongoing struggle against terrorist organizations. The related historical scholarship has focused on campaigns and experiences most similar, either in the form of the insurgency or the locale involved, but has almost exclusively discussed land power and ground operations. There has likewise been some examination of air power's role in small wars and irregular conflicts. A great deal of thinking and writing in this field is based on a foundation of historical research. Scholars and practitioners have written much less on the maritime elements of these same challenges.

In his book *Seapower: A Guide for the Twenty-First Century*, Geoffrey Till divided maritime forces at the beginning of the twenty-first century into two classes: the modern and postmodern navies. He defined modern navies as those built on late nineteenth- and twentieth-century ideals of blue-water fleet battles and power projection: designed for major power conflict and conventional in outlook. Today they focus on missions of nuclear deterrence and ballistic missile defense, sea control in blue water, highly kinetic visions of power projection, and good order at sea in home waters or areas of specific national interest. In Till's construction, postmodern navies are focused on sea control in littoral regions, expeditionary operations, a broad and cooperative vision of good order at sea, and maintenance of a maritime consensus.[4]

Yet the historical cases examined in these chapters demonstrate that much of the operational history of the United States' naval forces in the Age of Sail, or what might be termed the American premodern period, had a focus on the very areas Till identified as postmodern. Other examples, including naval operations in the Second Seminole War and the Greek Islands as well as those in the era immediately following the period examined in this book, including operations in Fiji, China, and Vietnam, support this conclusion.[5] Naval irregular warfare, rather than being a new development or a special element of recent, postmodern, naval conflict and peacetime responsibility, is actually quite old. The examples of the premodern in these chapters and the twenty-first century's postmodern, as described by Till, combine with examples of intervening "modern" years like

American naval interventions in the Caribbean, the Yangtze patrols of the 1920s, and riverine raiding of the Vietnam conflict. Together they begin to illustrate the fact that naval irregular warfare is not quite so irregular, but instead is a fundamental part of the entire operational history of the United States Navy and Marine Corps.[6]

A reappraisal of the U.S. Navy's formative years during the Age of Sail recognizes the role of naval irregular warfare as an element of the services' responsibility to the nation, both in wartime and during peace. Distinguishing this history as a fundamental component of naval activity offers a critical and expanded vision of the operational elements of American sea power. The more nuanced understanding this history—of why the U.S. Navy and Marine Corps were founded and how navies more generally operated across the spectrum of conflict—offers a correction to the self-image of naval forces exclusively as oceanic militaries required for decisive sea battle. It can also play a vital role in modern understanding of the maritime responsibilities of a powerful nation. If historians, strategists, and officers move beyond a view of naval strategy and operations as the dichotomy between guerre de course and guerre d'escadre, to include the concept of guerre de razzia and irregular warfare as another optional framework, it creates possibilities that allow them to address a spectrum of conflict, from peacetime to war. This greater awareness helps current and future naval leaders develop the full range of skills and insights needed to deal with irregular challenges: rising piracy, threats imbedded in maritime crime, and the increasing use of maritime measures short of war by competing regional powers, alongside traditional training for battle at sea and conventional power projection ashore.[7]

Strategists and historians have long divided naval operations into the binary categories of fleet battle or commerce raiding, known by their French labels of guerre de course and guerre d'escadre. For over a century, historians have laid this dual strategic construct across how they view the naval past. And yet, despite common acceptance of the construct and its application from Mahan to the modern day, there are many episodes in American and world naval history that do not fit comfortably inside these categories. As the past eight chapters demonstrate, there has been a third way in naval warfare and maritime operations. Maritime raiding and the execution of naval irregular warfare missions played a central part in the operational history of the early American Navy and Marine Corps. Just as the historical foundations of guerre de course and guerre d'escadre were established based on the naval history of the eighteenth and nineteenth centuries, these examples from the American experience in the Age of Sail demonstrate that the concept of guerre de razzia has not only intellectually sound reasoning behind it, but historical and practical credibility as well.

The legacy of Alfred Thayer Mahan casts a long shadow over naval strategy and theory, as well as the way historians approach the subject of naval affairs. His focus on the need to establish command of the sea, on the priority of guerre d'escadre and the historical examples of its importance, has guided many scholars in subsequent years. Some of these scholars have constructed a caricature of a man exclusively focused on large fleet battles and nothing else.[8] Yet other theorists and writers of Mahan's era, while they shared his belief in the importance of command, went further and cast a wider net by discussing a greater variety of types of naval operations. Sir Julian Corbett's writing on maritime expeditions and amphibious operations tends to be the common example contrasted with Mahan. However, modern scholars and practitioners who want to understand how sea power is used need to re-examine the role of maritime raiding and irregular warfare as well. A renewed engagement with Vice Admiral P. H. Colomb's more extensive study of maritime expeditions, raiding, and amphibious warfare serves as a launching point for a larger investigation of what naval forces do in both peace and war.[9] This refreshed look at the theory of naval strategy and operations, one that includes guerre de razzia, requires reinforcement by an examination of the history of naval irregular missions and operations in multiple eras.

In the late nineteenth century, less than a year after Mahan published *The Influence of Sea Power Upon History*, Colomb published *Naval Warfare: Its Ruling Principles Historically Treated* in Great Britain. Colomb examined much of the same history as Mahan, the vast majority of it focused on the Royal Navy. The British officer, however, discussed a wider range of naval operations than the American did, including a section scrutinizing the history of amphibious operations and maritime raiding. He broke these into two categories, operations to take and hold territory in one group and maritime raids of destruction, which he also called "cross-ravaging," in another.[10]

Colomb wrote from the perspective of a global hegemon, a naval force he had served in during his career in the mid to late nineteenth century, which recognized that naval responsibilities required more than a focus on the peer, or near-peer, competitor. By contrast, Mahan wrote *The Influence of Sea Power Upon History* for an audience in the United States, which was a cut-rate naval power. The American's view was aspirational, with historical analysis offered to help build the U.S. Navy into a maritime force capable of fighting at sea with the other Great Powers. In the era of their publication Mahan's work overshadowed Colomb. When Sir Julian Corbett picked up the mantle of British naval thought he also considered amphibious operations, but he primarily focused on Colomb's first category of missions to take and hold territory.[11] As a result, Colomb's inclusion of maritime raiding as an element of naval warfare and as an important part of naval history was overwhelmed by the blue-water focus that came to dominate the field as the world became obsessed with battle fleets and battleships.[12]

From the French thinkers of the Jeune École and Alfred Thayer Mahan to modern histories from Kenneth Hagan, George Baer, and others, this dichotomy of commerce raiding and battle fleets within naval strategy remains something of a dogma.[13] However, in 2003, the historian James Bradford returned to the operations and thinking of John Paul Jones and suggested that he may have had another concept in mind, a third way in naval warfare. Bradford spent an early portion of his scholarly career collecting, collating, transcribing, and editing the collected papers of Paul Jones. From this deep study of the early American captain, he identified the "alternative [school of naval strategy], *guerre de razzia* . . . first advocated by John Paul Jones in 1776."[14]

Bradford translated guerre de razzia, a phrase borrowed from the work of French colonial officers involved in mid-twentieth-century North African counterinsurgency, as "war by raiding." The concept has a long history in land warfare, including the colonial American experience in conflicts with Native Americans and in conflicts between European powers in the Americas.[15] Based on Paul Jones's correspondence and his operational record, Bradford connected it to the maritime world and identified it as "a style of warfare in which the main goal of operations is not the capture or destruction of the enemy's commerce as in *guerre de course*, or the defeat of his fleet as in *guerre d'escadre*, but the raiding of his coasts and colonies."[16] Bradford was the first to use guerre de razzia in a maritime context, but his work echoes Colomb's efforts a century earlier. Paul Jones's explicit pursuit of raiding operations and irregular missions as an element of American sea power demonstrated that the blue-water focus of the two traditional naval schools offered an incomplete framework.

Yet for guerre de razzia to be a legitimate naval concept it requires further historical scrutiny beyond Bradford's examination of Paul Jones's correspondence. Just as guerre de course and guerre d'escadre have discreet tactical and operational elements—capabilities, concepts, and skills that define their success and failure—a third school would require the same. These tactics and methods are best labeled as naval irregular warfare, and their existence and support for the concept of guerre de razzia are best examined through the study of naval irregular operations across multiple examples, beginning with the Age of Sail era of the U.S. Navy during which Paul Jones sailed.[17]

Over the course of the eight episodes examined in the previous chapters, these common concepts, capabilities, and skills have emerged. They align with the factors that Colomb highlighted as important to the success of raiding operations, as well as with Bradford's identification of John Paul Jones's thinking beyond conventional blue-water missions. The development of guerre de razzia as an element of early American sea power was not overtly theorized or discussed but was instead implicit and inherent to the operational and political challenges faced. But this is less reflective of its lesser or unimportant status than it is of an early American naval profession that had few explicit doctrinal concepts or categorical strategic teachings.[18]

Early American naval officers did not conceive of their operations using the strategic ideals and structure used by today's strategists and historians to describe naval operations in the modern world. While they did recognize that different skills and methods were involved, generally they did not appear to place too high a value on the differentiation between fleet operations and irregular ones. Instead, they pursued a more practical approach. By extension, while commerce raiding or fleet battle were used as the guiding ideas for American naval policy, guerre de razzia never was. In the nineteenth century, American squadrons and ships were built with an expectation that commerce raiding would be the predominant element of the American approach in war, but it was never exclusive in American strategy or combat operations. The examples Roosevelt used in his speech at the Naval Academy belie that fact, with the warship-against-warship combat of the War of 1812, Revolution, and Civil War playing a role in those conflicts. Likewise, in the twentieth century, even as American leaders adopted a policy of building for fleet battle, commerce raiding was still an element of how wars were fought. The Pacific submarine campaign in the Second World War demonstrates this fact. Effective naval strategy is about the mixing and balancing of different types of naval operations, rather than picking one exclusively. In this practical reality, guerre de razzia rises to its role as a third way of naval warfare, one of the elements to be mixed and balanced in effective strategy.

But in modern interpretations of the history, both historians and naval professionals continue to place great stock in the importance of these categories and labels. They place the highest value on blue-water operations and fleet battle. Some naval historians have suggested that recording and explaining the battles between great fleets is the raison d'être of naval history.[19] They place their priority squarely on battle at sea between opposing fleets, and assert that combat at sea is "the touchstone of naval history" because it demonstrates "the extent to which navies achieved the reality of their own self-images."[20] Other historians may find this definition too narrow because the approach begs the question of whether "self-image" is the best way to frame historical study.

How individuals or organizations conceive of themselves does not always match with reality or historical accuracy. As Arthur Marder described in his chapter "The Influence of History on Sea Power," naval forces are not always adept at accepting accurate depictions of their own image or their own history.[21]

Whether through an anti-intellectualism or a particular operational bias,[22] the economic and national policy implications of building fleets,[23] or simple self-interest and professional preservation,[24] relying on that self-image to define how and what naval historians should study is an unreliable measure. It provides an unbalanced view of the past.

The conventional narratives and traditional approaches to naval history have laid an important foundation for the study of the early American Navy and Marine Corps: detailing what politicians, senior officers, and civilian officials were doing during the period, constructing a navy prepared for war and waging large-scale blue-water operations.[25] And some of the literature and research on the maritime past does reference naval irregular warfare and sometimes specifically lists operations studied in the chapters of this book. However, this history, in keeping the prevailing focus on oceanic operations, tends to mention the operations in passing. It offers a page or a paragraph as a bridge to move from one war to another. The events are included with little comparison or connection to other naval themes or to other irregular examples, resulting in far less scrutiny or analytical depth when contrasted with the study of blue-water battle. This approach remains in keeping with the norms of the field but offers limited insight for a balanced understanding of the naval operations of the period. A handful of efforts offer a good foundation for historians looking to expand our understanding of a wider view of naval operational history, and the role of naval irregular warfare.[26] Yet there is much more history still to be researched and analyzed in testing the concept of guerre de razzia and its place in the strategic frameworks that are still used today.

Notes

Abbreviations

ADM	Admiralty Files, The National Archives, Kew, UK
ASP NA	American State Papers, Naval Affairs, Library of Congress, Washington, D.C.
Barbary	*Naval Documents Related to the United States Wars with the Barbary Powers* (6 vols.)
DANFS	*Dictionary of American Naval Fighting Ships*
DHCNR	*Documentary History of the Campaign on the Niagara Frontier, 1813* (4 vols.)
Documents	*Documents, Official and Unofficial, Relating to the Case of the Destruction of the Frigate Philadelphia at Tripoli, on the 16th February, 1804*
EPP	Edward Preble Papers, Library of Congress, Washington, D.C.
Green Diary	*Diary of Ezra Green, M.D., Surgeon on Board the Continental Ship-of-War "Ranger," under Capt. John Paul Jones, from November 1, 1777 to September 27, 1778*
Hopkins	*The Letter Book of Esek Hopkins, Commander-in-Chief of the United States Navy, 1775–1777*
JPJC	*John Paul Jones Commemoration at Annapolis, April 24, 1906*
JPJ Papers	*The Papers of John Paul Jones*
JPJ&R	*John Paul Jones and the Ranger: Portsmouth, New Hampshire, July 12–November 1, 1777, and the Log of the Ranger, November 1, 1777–May 18, 1778*
LAC RG 8	Library and Archives of Canada (Ottawa, Ontario), Record Group 8
M #, Roll #	Microfilm Volume, Microfilm Roll
Memoir	*John Paul Jones' Memoir of the American Revolution: Presented to King Louis XVI of France*
NARA	U.S. National Archives and Records Administration, Washington, D.C.
NASP	*The New American State Papers, Naval Affairs*
NDAR	*Naval Documents of the American Revolution* (vols. 6, 7, 12)
NWIP	"Our Navy and the West Indian Pirates: A Documentary History" (12 parts, U.S. Naval Institute *Proceedings*)
Parker Journal	Midshipman James L. Parker, *Journal of a Cruise Around the World*

PCM   *Minutes of the Proceedings of the Courts of Inquiry and Court Martial, in Relation to Captain David Porter*

RG 45   Record Group 45, U.S. National Archives and Records Administration, Washington, D.C.

SAL   *A Century of Lawmaking for a New Nation: U.S. Congressional Documents and Debates, 1774–1875*, Statutes at Large, Library of Congress, Washington, D.C.

W1812   *The Naval War of 1812: A Documentary History* (3 vols.)

Introduction

1. "Summary" and "Order of Rear-Admiral Sands, 24 April 1906," in C. W. Stewart, *John Paul Jones Commemoration at Annapolis*, address of Theodore Roosevelt, 11, 204–5. Martelle, *The Admiral and the Ambassador*, 210–11, 244–45, 273.
2. Roosevelt, *John Paul Jones*, 19.
3. Ibid., 16–18. Final movement to the crypt occurred in 1913. Martelle, *The Admiral and the Ambassador*, 273.
4. Hagan, *American Gunboat Diplomacy and the Old Navy*, 11.
5. Mahan, *Influence of Sea Power upon History*, 394–96.
6. James Bradford first suggested the application of the land power concept of guerre de razzia to the sea after his examination of the operational history and correspondence of John Paul Jones. While the concept of raiding and irregular warfare has a large body of literature referencing land power and armies, there is far less consideration of naval operations. Bradford, "John Paul Jones and Guerre de Razzia," 2.
7. Cable, *Gunboat Diplomacy*, 15–21.
8. Thornton, *Asymmetric Warfare*, 1–5; Schubert, *Other Than War*, 5–6; Mattis and Hoffman, "Future Warfare," 18–19.
9. Till, *Asia's Naval Expansion*, 163.
10. Colomb, *Naval Warfare*, 22–25, 251–52.
11. For excellent studies on the Jeffersonian gunboats see Tucker, *The Jeffersonian Gunboat Navy*, and Smith, *For the Purposes of Defense*.
12. Armstrong, "D— All of the Above," 13–17.
13. U.S. Department of the Navy, *Naval Operations Concept 2010: Implementing the Maritime Strategy*, https://fas.org/irp/doddir/navy/noc2010.pdf.
14. Symonds, *Navalists and Antinavalists*, 222–25; Slaughter, "Genesis of the U.S. Navy," in Bradford, *America, Seapower, and the World*, 26–41.
15. *Examination Papers: United States Naval Academy, 1881–82.*
16. McKee, *A Gentlemanly and Honorable Profession*, 155–56, 215.
17. Ethan Rafuse makes this case in his *Journal of Military History* article "'Little Phil,' a 'Bad Old Man,' and the "Gray Ghost': Hybrid Warfare and the Fight for the Shenandoah Valley, August–November 1864," 776–79.
18. Till, *Seapower*, 35–41.

Chapter 1. John Paul Jones and the Birth of American Naval Irregular Warfare

1. Armstrong, "An Act of War on the Eve of Revolution," 43–48.
2. Testimony of Bartholomew Cheever, Minutes of a Court Martial Assembled on board His Majesty's Ship Centaur in Portsmouth Harbor, 14 October 1772, in *William R. Staples' The Documentary History of the Destruction of the Gaspee*, introduced and

supplemented by Richard M. Deasy, 135. Statement of Bartholomew Cheever, 10 June 1772, Staples, *Documentary History*, 11.

3. Exhibit A, Statement of W. Dudingston, Minutes of a Court Martial, 14 October 1772, in Deasy, *William R. Staples*, 137; statement of Ephraim Bowen, 29 August 1839; in Staples, *Documentary History*, 8–9.

4. Statement of John Mawney, 1826, in Staples, *Documentary History*, 9–10; Exhibit A, Statement of W. Dudingston, Minutes of a Court Martial, 14 October 1772, in Deasy, *William R. Staples*, 138.

5. Darrius Sessions to J. Wanton, 21 March 1772, W. Dudingston to J. Wanton, 23 March 1772, in Staples, *Documentary History*, 1, 4.

6. Park, *The Burning of His Majesty's Schooner Gaspee*, 15–19.

7. *The Boston Chronicle*, 24 July 1769, vol. 2, no. 30, 236; Armstrong, "An Act of War on the Eve of Revolution," 43–48.

8. Paullin, *The Navy of the American Revolution*; McGrath, *Give Me a Fast Ship*.

9. Paul Jones, *Memoir of the American Revolution*.

10. MacDermott, "Notes: Father of the American Navy," 71–72.

11. Grant, "The Resurrection of John Paul Jones," 52–57.

12. Davis, *Some Facts about John Paul Jones*, 13, 16.

13. Bradford, *Command under Sail*, 21–22.

14. Lorenz, *John Paul Jones: Fighter for Freedom and Glory*, 65–67.

15. Bradford, "John Paul Jones: Honor and Professionalism," 22–23.

16. Paul Jones to Morris, 17 October 1776, *NDAR* 6:1302; Rider, *Valour Fore & Aft*, 88.

17. Paul Jones to Morris, 17 October 1776, *NDAR* 6:1302; Morison, *Jones*, 61, 64–65; Rider, *Valour Fore & Aft*, 88–89.

18. Hopkins to Paul Jones, 22 November 1776, *Letter Book of Esek Hopkins*, 97–88.

19. Hopkins to Hancock, 24 October 1776, *Letter Book of Esek Hopkins*, 89–90.

20. Paul Jones to Morris, 17 October 1776, *NDAR* 6:1302.

21. Paul Jones to Continental Marine Committee, 30 October 1776, *NDAR* 6:1457. Paul Jones to Hacker, 1 November 1776, *NDAR* 7:6.

22. Paul Jones to Continental Marine Committee, 12 November 1776, *NDAR* 7:111.

23. Paul Jones to Continental Marine Committee, 12 January 1777, *NDAR* 7:935–36.

24. Paul Jones to Morris, 17 October 1776, *NDAR* 7:1303–4. Morison, *Jones*, 14–15.

25. Morris to Paul Jones, 5 February 1777, *NDAR* 7:1109–11.

26. Ibid., 1110.

27. Morris to Hopkins, 5 February 1777, *NDAR* 7:1111–12. Paul Jones to Hopkins, 28 February 1777, *NDAR* 7:1315.

28. Hopkins to Marine Committee, 14 February 1776, *Letter Book of Esek Hopkins*, 121–22. Hopkins to Morris, 28 February 1777, *Letter Book of Esek Hopkins*, 125–26. Hopkins to the Marine Committee, 28 February 1777, *Letter Book of Esek Hopkins*, 126–27.

29. "John Langdon," *Biographical Directory of the United States Congress*. Chapelle, Naval appropriation act of November 1776 in *The History of the American Sailing Navy*, 79–86.

30. Langdon to Hamilton, 6 November 1776, *NDAR* 7:58–59.

31. Langdon to Whipple, 4 December 1776, *NDAR* 7:362–63.

32. Marine Committee to Paul Jones, 18 June 1777, *JPJ&R*, 15.

33. Paul Jones to Langdon, 11 September 1777, *JPJ&R*, 41.

34. Joseph Sawtelle, "The Ranger, 1777," *JPJ&R*, 5.
35. Paul Jones to Continental Marine Committee, 10 December 1777, *JPJ&R*, 65.
36. Green, *Diary*, 18.
37. Ibid. Log of the Continental Navy Ship *Ranger*, 26 November 1777, *JPJ&R*, 57–58.
38. Log of Ranger, 2 December 1777, *JPJ&R*, 61–62.
39. Paul Jones to Wendell, 11 December 1777, *JPJ Papers*, Roll 1, 219.
40. Paul Jones to Continental Marine Committee, 10 December 1777, *JPJ Papers*, Roll 1, 216.
41. Paul Jones to Whipple, 11 December 1777, *JPJ Papers*, Roll 1, 220.
42. Paul Jones to Continental Marine Committee, 10 December 1777, *JPJ Papers*, Roll 1, 216.
43. Franklin and Deane to Paul Jones, 16 January 1778, *The Revolutionary Diplomatic Correspondence of the United States*, 2: 471–72.
44. Continental Marine Committee to Paul Jones, *JPJ Papers*, Roll 2, 233. Bradford, "Guerre de Razzia," 6.
45. Franklin and Deane to Paul Jones, 16 January 1778, *Revolutionary Diplomatic Correspondence*, 2:471–72.
46. Simpson, Hall, and Cullam to Paul Jones, 14 February 1778, *JPJ Papers*, Roll 2, 251. Paul Jones to The Eastern Navy Board, 23 February 1778, *JPJ Papers*, Roll 2, 256. Parke to Paul Jones, 19 February 1778, *JPJ Papers*, Roll 2, 253.
47. Green, *Diary*, 21.
48. Log of Ranger, 14 April 1778, *JPJ&R*, 136. Green, *Diary*, 19.
49. Log of Ranger, 17 April 1778, *JPJ&R*, 138. Paul Jones to American Commissioners, 27 May 1778, *JPJ Papers*, Roll 2, 310.
50. John Paul Jones to President of Continental Congress, 7 December 1779, *JPJ Papers*, Roll 5, 910. De Koven, *The Life and Letters of John Paul Jones*, 1:280.
51. Morison, *Jones*, 135–37. Paul Jones to American Commissioners, 27 May 1778, *JPJ Papers*, Roll 2, 310.
52. Log of Ranger, 17 April 1778, *JPJ&R*, 139. Paul Jones to American Commissioners, 27 May 1778, *JPJ Papers*, Roll 2, 310.
53. Paul Jones to American Commissioners, in Sherburne, *Life and Character of the Chevalier John Paul Jones*, 44. *Cumberland Chronicle Extraordinary*, 23 April 1778, in *Paul Jones: His Exploits*, edited by Seitz, 6–7.
54. Paul Jones to American Commissioners, 27 May 1778, *JPJ Papers*, Roll 2, 310.
55. 20–21 April 1778, Log of Ranger, *JPJ&R*, 140–41. The ship was HMS *Drake* which *Ranger* would encounter again.
56. Green, *Diary*, 21.
57. Paul Jones to American Commissioners, 27 May 1778, *JPJ Papers*, Roll 2, 310.
58. Ibid. Green, *Diary*, 22. Accusation of drunken mate in John Paul Jones, "Journal of Paul Jones," *Niles Weekly Register* 2 (13 June 1812): 249–51.
59. Green, *Diary*, 21.
60. Log of Ranger, 21–23 April 1778, *JPJ&R*, 141–42.
61. Bowen-Hassell, Conrad, and Hayes, *Sea Raiders of the American Revolution*.
62. "Examination of David Freeman," Whitehaven, England, 24 April 1777, *NDAR* 12:596.
63. Paul Jones to American Commissioners, 27 May 1778, *JPJ Papers*, Roll 2, 310.

64. W. Brownrigg and Henry Ellison, Justices of the Peace, "Sequel of the Proceedings of Capt Paul Jones and his Crew at Whitehaven," 24 April 1777, *NDAR* 12:597.

65. Paul Jones, *Memoir of the American Revolution*, 18. "Sequel of the Proceeding," *NDAR* 12:597. Paul Jones to American Commissioners, 27 May 1778, *JPJ Papers*, Roll 2, 310.

66. "Sequel of the Proceeding," *NDAR* 12:597.

67. Paul Jones to American Commissioners, 27 May 1778, *JPJ Papers*, Roll 2, 310.

68. Ibid. "Sequel of the Proceeding," *NDAR* 12:597. "Examination of David Freeman," Whitehaven, England, 24 April 1777. Log of Ranger, 23 April 1778, *JPJ&R*, 142.

69. *The Morning Post and Daily Advertiser*, 28 April 1778, *JPJ&R*, 209.

70. Paul Jones to American Commissioners, 27 May 1778, *JPJ Papers*, Roll 2, 310.

71. Ibid. Log of Ranger, 23 April 1778, *JPJ&R*, 142.

72. *The Morning Post and Daily Advertiser*, 28 April 1778. Log of Ranger, 23 April 1778, *JPJ&R*, 142.

73. Paul Jones to American Commissioners, 27 May 1778, *JPJ Papers*, Roll 2, 310.

74. "Sequel of the Proceeding," *NDAR* 12:597.

75. Paul, *The Scots Peerage*, 7:520–21.

76. Paul Jones to Hamilton, 8 May 1778, *NDAR* 12:675. Paul Jones, *Memoir of the American Revolution*, 19.

77. Paul Jones, *Memoir of the American Revolution*, 19. Morison, *Jones*, 144.

78. *The Gazetteer and New Daily Advertiser*, 1 May 1778, *NDAR* 12:643.

79. Paul Jones, *Memoir of the American Revolution*, 20. In a strange ending to the story of the Selkirk silver, Paul Jones wrote a letter directly to Lady Hamilton apologizing for the entire episode. In the letter he wrote as if the two were old friends and made an awkward attempt to convince her of the importance and mercy of the American cause. On the advice of John Ross, he used his own money to buy the plate from his crew and then returned it to the Selkirk estate in an attempted display of honor. Paul Jones to Hamilton, 8 May 1778, *NDAR* 12:675–76. John Ross to John Paul Jones, 17 May 1778, *NDAR* 12:704. Paul Jones, *Memoir of the American Revolution*, 21.

80. Gell to Stevens, 24 April 1778, *NDAR* 12:594.

81. Stevens to Fraser, 2 May 1778, *NDAR* 12:647.

82. Log of Ranger, 23 April 1778, *JPJ&R*, 143. Paul Jones, *Memoir of the American Revolution*, 20.

83. "Extract of a Letter from Port-Glasgow, 27 April," *Daily Advertiser* (London), 4 May 1778, No. 14773. Log of Ranger, 24 April 1778, *JPJ&R*, 144.

84. Paul Jones, *Memoir of the American Revolution*, 20.

85. Paul Jones to American Commissioners, 27 May 1778, *JPJ Papers*, Roll 2, 310. Friday, 24 June, Green, *Diary*, 23.

86. Paul Jones to American Commissioners, 27 May 1778, *JPJ Papers*, Roll 2, 310.

87. Ibid.

88. Friday, 24 June, Green, *Diary*, 23. Log of Ranger, 24 April 1778, *JPJ&R*, 144.

89. Friday, 24 June, and Monday, 4 May, Green, *Diary*, 23, 25.

90. Paul Jones to American Commissioners, 27 May 1778, *JPJ Papers*, Roll 2, 310.

91. Log of Ranger, 24–26 April 1778, *JPJ&R*, 144.

92. Paul Jones, *Memoir of the American Revolution*, 21. Saturday, 25 June, Green, *Diary*, 25. Paul Jones to American Commissioners, 27 May 1778, *JPJ Papers*, Roll 2, 310.

93. Crawford to Stevens, 26 April 1778, *NDAR* 12:604–5.

94. Paul Jones, *Memoir of the American Revolution*, 21. While *Ranger* and the prizes did not fall in with any other British ships, an altercation did occur between Paul Jones and Thomas Simpson, his first lieutenant, whom he assigned as prize master of *Drake*. After a case of mistaken, or ignored, signals in the final days of the voyage back to Brest, *Drake* and *Ranger* were separated. Paul Jones insisted it was willful insubordination, the problems with his officers and crew coming to a head, placed Simpson under arrest, and preferred charges for a court-martial.
95. American Commissioners to Paul Jones, 25 May 1778, *NDAR* 12:747. Franklin to Paul Jones, 27 May 1778, in *JPJ Papers*, Roll 2, 316.
96. *The Morning and Daily Advertiser*, 28 April 1778, *JPJ&R*, 211.
97. *The Morning and Daily Advertiser*, 5 May 1778, *JPJ&R*, 212.
98. *Morning Chronicle and London Advertiser*, 8 May 1778, *JPJ&R*, 214. *The Morning and Daily Advertiser*, 5 May 1778, *JPJ&R*, 212.
99. *London Evening Post*, April 25–28, 1778, Issue 8754.
100. *The Morning and Daily Advertiser*, 28 April 1778, *JPJ&R*, 211.
101. "Extract of a Letter from Whitehaven, May 16th," *Lloyd's Evening Post*, 22–25 May 1778, *NDAR* 12:699–700.
102. *St. James's Chronicle/British Evening Post* (London), June 6–9, 1778, Issue 2679.
103. Paul Jones to Twitcher, *London Evening Post*, in *Paul Jones: His Exploits*, edited by Seitz, 75.
104. *London Evening Post*, 28 September 1779, Issue 8966.
105. Gilpin, *Observations*, 2:105–6. Laughton, *Studies in Naval History: Biographies*, 363.
106. Paul Jones to American Commissioners, 27 May 1778, *JPJ Papers*, Roll 2, 310.
107. John Paul Jones to President of Continental Congress, 7 December 1779, *JPJ Papers*, Roll 5, 910.
108. Allen, *Battles of the British Navy*, 1:206–7. Laughton, *Studies in Naval History: Biographies*, 361. As with many historical naval engagements, an element of luck likely also had a hand in the different results for Thurot and Paul Jones.
109. Morison, *Jones*, 142. Seconded by Bradford, "Guerre de Razzia," 7.
110. Paul Jones to American Commissioners, 27 May 1778, *JPJ Papers*, Roll 2, 310.

Chapter 2. Wars Done by Halves

1. Chapelle, *The History of the American Sailing Navy*, 80. Symonds, *Navalists and Antinavalists*, 17–37. Constitution of the United States, Article I, Section 8, National Archives and Records Administration (NARA).
2. Nash, *Forgotten Wars*, 34–35. Sharrer, "The Search for a Naval Policy," in Hagan, *In Peace and War*, 30–31. Smith, *Patriarch*, 180.
3. "Act to Provide a Naval Armament," 27 March 1794, in *SAL*, 3rd Congress, 1st Session, 350–51. "Treaty of Peace and Amity, Signed at Algiers, September 5th, 1795," in Miller, *Treaties and Other International Acts*, vol. 2.
4. Washington's Farewell Address, *The Papers of George Washington Digital Edition*. DeConde, *The Quasi-War*, 9–10. Crawford and Hughes, *The Reestablishment of the Navy*, 6–7.
5. George Washington to Congress, 15 March 1796, and Bingham to Senate, 17 March 1796, *ASP NA* 1:25. Sharrer, "Naval Policy," 33. Smith, *Patriarch*, 258.
6. Grant, *John Adams*, 388–91. DeConde, *The Quasi-War*, 16–17, 46–48.
7. "Act to Provide an Additional Armament for the further protection of the trade

of the United States and for other purposes," 27 April 1798, *SAL* 1:552. "An Act to establish an Executive department, to be denominated the Department of the Navy," 30 April 1798, *SAL*, 5th Congress, 2nd Session, 553–54. Crawford and Hughes, *The Reestablishment of the Navy*, 8. Nash, *Forgotten Wars*, 52–53.

8. Palmer, *Stoddert's War*, 14–16. Nash, *Forgotten Wars*, 55–57. "An Act to authorize the President of the United States to cause to be purchased, or built, a number of small vessels to be equipped as gallies or otherwise," 4 May 1798, *SAL*, 5th Congress, 2nd Session, 556. "An Act to amend the act, instituted 'An act to authorize the President of the United States to cause to be purchased or built, a number of vessels, to be equipped as gallies or otherwise,'" 22 June 1798, *SAL*, 5th Congress, 2nd Session, 569. "An Act to authorize the defense of Merchant Vessels of the United States against French depredations," 25 June 1798, *SAL*, 5th Congress, 2nd Session, 572–73.

9. "An Act more effectually to protect the Commerce and Coasts of the United States," 28 May 1798, *SAL*, 5th Congress, 2nd Session, 561.

10. "An Act further to protect the Commerce of the United States," 9 July 1798, *SAL*, 5th Congress, 2nd Session, 578. "An Act to Make a further appropriation for the additional Naval Armament," *SAL*, 5th Congress, 2nd Session, 608–9.

11. *Dictionary of American Naval Fighting Ships* (DANFS), *Ganges*. Leiner, *Millions for Defense*, 26.

12. In the interests of modern understanding, this chapter will refer to the eastern portion of the island of Hispaniola as Haiti. During the era studied, however, it was more commonly referred to by its French colonial name of Saint Domingue until the nation of Haiti was created in 1804.

13. Palmer, *Stoddert's War*, 233–34. Thomas Truxton to Benjamin Stoddert, January 1799, *Naval Documents Related to the Quasi-War with France*, 2:356–58. Mattis and Hoffman, "Hybrid Warfare," 18.

14. "An Act more effectually to protect the Commerce and Coasts of the United States," 28 May 1798, *SAL* 1, 561.

15. Stoddert to Commanders of United States Armed Vessels, 10 July 1798, *Naval Documents Related to the Quasi-War with France*, 1:187.

16. DANFS—*Ganges, Delaware II*.

17. Stoddert to Adams, 30 July 1798, *Naval Documents Related to the Quasi-War with France*, 1:255–56.

18. Allen, *Our Naval War with France*, 67–69. Nash, *Forgotten Wars*, 94.

19. Truxton to Stoddert, 27 October 1798, *Naval Documents Related to the Quasi-War with France*, 1:566–68. Palmer, *Stoddert's War*, 238.

20. Stoddert to Talbot, 14 November 1799, *Naval Documents Related to the Quasi-War with France*, 4:391–92. Stoddert to Talbot, 4 December 1799, *Naval Documents Related to the Quasi-War with France*, 4:480–481. Palmer, *Stoddert's War*, 164–65. Girard, *Toussaint Louverture*, 177–78, 186. Dubois, *A Colony of Citizens*, 306–7. Port-au-Prince's name was changed to Port-Republicain in 1793 and reverted to its prior name after Haiti declared independence. In the interest of modern understanding, this chapter will use Port-au-Prince.

21. Protest of Hart, Holland, and Done, Seamen of the American Schooner Mary, 1 January 1800, *Naval Documents Related to the Quasi-War with France*, 5:5–6. Journal of William Maley, 1 January 1800, *Naval Documents Related to the Quasi-War with France*, 5:4. Cape François, or Cap Français, is known today as Cap Haitian.

22. DANFS—*Experiment I*. Chapelle, *The History of the American Sailing Navy*, 145–46.

23. Palmer, *Stoddert's War*, 166, 168–69.

24. Open letter of Silas Talbot to the American public and Merchants, 12 February 1800, *Naval Documents Related to the Quasi-War with France*, 5:208–9. Girard, *Toussaint Louverture*, 186–87. Allen, *Our Naval War with France*, 120.

25. Journal of Maley, 1 January 1800, *Naval Documents Related to the Quasi-War with France*, 5:4.

26. Stephens to Talbot, 2 January 1800, *Naval Documents Related to the Quasi-War with France*, 5:1.

27. Journal of Maley, 1 January 1800, *Naval Documents Related to the Quasi-War with France*, 5:4.

28. Stephens to Talbot, 2 January 1800, *Naval Documents Related to the Quasi-War with France*, 5:2. Protest of Hart, Holland, and Done, 1 January 1800, *Naval Documents Related to the Quasi-War with France*, 5:5–6.

29. Protest of Hart, Holland, and Done, 1 January 1800, *Naval Documents Related to the Quasi-War with France*, 5:5. Stephens to Talbot, 2 January 1800, *Naval Documents Related to the Quasi-War with France*, 5:2.

30. Stephens to Talbot, 2 January 1800, *Naval Documents Related to the Quasi-War with France*, 5:2.

31. Protest of Hart, Holland, and Done, 1 January 1800, *Naval Documents Related to the Quasi-War with France*, 5:5. Stephens to Talbot, 2 January 1800, *Naval Documents Related to the Quasi-War with France*, 5:2.

32. Journal of Maley, 1 January 1800, *Naval Documents Related to the Quasi-War with France*, 5:4. Stephens to Talbot, 2 January 1800, *Naval Documents Related to the Quasi-War with France*, 5:3.

33. Adm. David Dixon Porter made a claim in his published memoir of his father that Maley had been a coward during the engagement and Porter assumed command and saved *Experiment*. A number of biographers repeated the story based on the Porter family recollection. However, Michael Palmer used close analysis of the contemporary source documents to demonstrate that Porter was almost assuredly mistaken and confused with another incident on board *Experiment* the following February. Porter, *Memoir of Commodore David Porter of the United States Navy*, 29–31. Story repeated in Fowler, *Jack Tars and Commodores*, 51–52; and Long, *Nothing Too Daring*, 12–13. Palmer, *Stoddert's War*, 167–68.

34. Silas Talbot to David Porter, 3 March 1800, *Naval Documents Related to the Quasi-War with France*, 5:263–264.

35. Journal of the French Privateer Ester, 30 April–1 May, 1800, *Naval Documents Related to the Quasi-War with France*, 5:468. Journal of Isaac Hull, 3 May 1800, *Naval Documents Related to the Quasi-War with France*, 5:474. Journal of Ester, 2 May 1800, *Naval Documents Related to the Quasi-War with France*, 5:471.

36. Norton, *Captains Contentious*, 48, 52. Fowler, *Silas Talbot*, 46–48. Palmer, *Stoddert's War*, 175.

37. Palmer, *Stoddert's War*, 175. Allen, *Our Naval War with France*, 182–83. Fowler, *Silas Talbot*, 48.

38. Hull Journal, 5 May 1800, *Naval Documents Related to the Quasi-War with France*, 5:481. Hull Journal, 8 May 1800, *Naval Documents Related to the Quasi-War with France*, 5:491. Palmer, *Stoddert's War*, 176.

39. Journal of Constitution, 9 May 1800, *Naval Documents Related to the Quasi-War with France*, 5:495–96.

40. Hull Journal, 9 May 1800, *Naval Documents Related to the Quasi-War with France*, 5:495–96.

41. Hull Journal, 10 May 1800, *Naval Documents Related to the Quasi-War with France*, 5:499.

42. Deposition of Thomas Sandford, 19 May 1800, *Naval Documents Related to the Quasi-War with France*, 5:501. Palmer, *Stoddert's War*, 177.

43. Journal of Constitution, 11 May 1800, *Naval Documents Related to the Quasi-War with France*, 5:505. Deposition of Thomas Sandford, 19 May 1800, *Naval Documents Related to the Quasi-War with France*, 5:501.

44. Hull Journal, 11 May 1800, *Naval Documents Related to the Quasi-War with France*, 5:505–6. Silas Talbot to Benjamin Stoddert, 12 May 1800, *Naval Documents Related to the Quasi-War with France*, 5:503.

45. Hull papers—Account of Expedition in harbor of Porto Plata, 11 May 1800, in *Naval Documents Related to the Quasi-War with France*, 5:504–5.

46. Daniel Carmick to a friend in Philadelphia, 12 May 1800, *Naval Documents Related to the Quasi-War with France*, 5:500. Allen, *Our Naval War with France*, 183–84.

47. Palmer, *Stoddert's War*, 178. Allen, *Our Naval War with France*, 186.

48. Smith, *Marines in the Frigate Navy*, 9.

49. Daniel Carmick to a friend in Philadelphia, 12 May 1800, *Naval Documents Related to the Quasi-War with France*, 5:501.

50. Silas Talbot to Benjamin Stoddert, 12 May 1800, in *Naval Documents Related to the Quasi-War with France*, 5:503–4.

51. Benjamin Stoddert to Richard Harrison, 14 July 1800, *Naval Documents Related to the Quasi-War with France*, 6:150. Alexander Murray to Stoddert, 31 July 1800, *Naval Documents Related to the Quasi-War with France*, 6:210–11. Stoddert to Talbot, 5 September 1800, *Naval Documents Related to the Quasi-War with France*, 6:320. Palmer, *Stoddert's War*, 181–82. DeConde, *The Quasi-War*, 255–56.

52. See Perrett, *Gunboat!*; Preston, *Send A Gunboat*; Corbett, *Some Principles of Maritime Strategy*.

Chapter 3. Intrepid and Irregular Warfare on the Barbary Coast

1. Leiner, *The End of Barbary Terror*, 1–2.

2. Allen, *Our Navy and the Barbary Corsairs*, 90. Tucker, *Dawn Like Thunder*, 228.

3. Fowler, "The Navy's Barbary War Crucible," 55–58.

4. "Biographical Memoir of Commodore Dale," 514.

5. Fowler, "Navy's Barbary War Crucible," 55–58.

6. William Eaton to James Cathcart, 4 August 1802, Area Files of the Naval Records Collection, 1775–1910, National Archives Microfilm M 625. Wheelan, *Jefferson's War*, 156–57.

7. Tucker, *Dawn Like Thunder*; Whipple, *To the Shores of Tripoli*. Recent examples include: Kilmeade and Yaeger, *Thomas Jefferson and the Tripoli Pirates*; Wheelan, *Jefferson's War*; and Hitchens, "Jefferson Versus the Muslim Pirates," *Arguably*, 12–20. For a book focused on the economic rather than the theological drivers of the conflict see Lambert, *The Barbary Wars*.

8. Bainbridge to Smith, 1 November 1803, in *Barbary*, 3:171–73.

9. Ibid.

10. Letter from Officers of *Philadelphia* to William Bainbridge, Loss of the Frigate Philadelphia, 20 March 1804, *ASP NA*, 1:123.

11. Bainbridge to Smith, 1 November 1803, in *Barbary*, 3:171–73.

12. Extract from Diary of Edward Preble, 27 November 1803, in *Barbary*, 3:240.

13. Report of Senator Hayne, "On Claim of the Officers and Crew of the ketch Intrepid to prize money for the destruction of the frigate Philadelphia, at Tripoli, in 1804," 9 January 1828, *ASP NA*, 3:122–24. Edward Preble, dispatches, in *Documents, Official and Unofficial*, compiled by De Selding, 16.

14. William Bainbridge to Edward Preble, 5 December 1803, *Barbary*, 3:253–54. Preble to Robert Smith, 10 December 1803, *Barbary*, 3:258. Tucker, *Dawn Like Thunder*, 261–62. Pratt, *Preble's Boys*, 21. *Allegiance* was also known by *Meriam*, her American name prior to capture by the British.

15. Edward Preble to Stephen Decatur, 10 December 1803, *Barbary*, 3:260. Edward Preble Memoranda, 17 December 1803, *Barbary*, 3:278. Extract from log book kept by Sailing Master Nathaniel Haraden, U.S. Navy, 23 December 1803, *Barbary*, 3:288.

16. DANFS—*Intrepid*.

17. Tucker, *Stephen Decatur*, 42–43. Edward Preble to Hethcote J. Reed, 23 December 1803, in *Barbary*, 3:288–89.

18. Statement of Salvatore Catalano, 2 February 1804, *Barbary*, 4:181. Whipple, *To the Shores of Tripoli*, 131.

19. Edward Preble to Tobias Lear, 31 January 1804 *Barbary*, 3:378. For details of the court filings see "Documents concerning the participation of the ketch *Mastico* in the capture of the *Philadelphia*," *Barbary*, 4:180–81. Purchase data from *Register of Officer Personnel, United States Navy and Marine Corps, and Ships' Data*, edited by Knox, 74.

20. Preble to Lear, 17 January 1804, in *Barbary*, 3:340.

21. Report of Senator Hayne, "On Claim of the Officers and Crew of the ketch Intrepid." Tucker, *Stephen Decatur*, 43–44. Tucker, *Dawn Like Thunder*, 262–63. Both Tuckers cast doubt on Decatur's role in the planning, but Charles Stewart was quite clear that Decatur was the first of Preble's lieutenants to suggest a raid into the harbor: Charles Stewart to Mrs. Stephen Decatur, 12 December 1826, in *Barbary*, 3:426–27. Decatur's role as initial planner instead of Stewart supported in Allison, *Stephen Decatur*, 45–46.

22. Edward Preble to Stephen Decatur, 31 January 1804, *Documents, Official and Unofficial*, compiled by De Selding, 13.

23. DANFS—*Syren*. Edward Preble to Charles Stewart, 31 January 1804, *Barbary*, 3:375–76.

24. Tucker, *Dawn Like Thunder*, 271–72. Lewis Heermann, affidavit, 26 April 1828, *Documents, Official and Unofficial*, compiled by De Selding, 27–28.

25. Tucker, *Stephen Decatur*, 49–50. Bainbridge to Preble, 16 February 1804, *Barbary*, 3:410.

26. Certificate of Salvatore Catalano, 29 December 1825, *Documents, Official and Unofficial*, compiled by De Selding, 23.

27. Quoted in C. W. Goldsborough to Michael Hoffman, 10 March 1828, *Documents, Official and Unofficial*, compiled by De Selding, 25. Stephen Decatur to Edward Preble, 17 February 1804, *Documents, Official and Unofficial*, compiled by De Selding, 14.

28. Certificate of Salvatore Catalano, 29 December 1825, *Documents, Official and Unofficial*, compiled by De Selding, 23.

29. Decatur to Preble, 17 February 1804, *Documents, Official and Unofficial*, compiled by De Selding, 14. Tucker, *Stephen Decatur*, 51–55.

30. Robert Smith to Edward Preble, 22 May 1804, *Barbary*, 3:427. Ralph Izard to Mrs. Izard, 20 February 1804, *Barbary*, 3:416–17. James Cathcart to Edward Preble, 19 February 1804, and Edward Preble to James Cathcart, 19 February 1804, *Barbary*, 3:435–38. Tucker, *Dawn Like Thunder*, 290–91.

31. "Naval Operations against Tripoli," Thomas Jefferson to Congress, 20 February 1805, *Barbary*, 4:293–94. Tucker, *Dawn Like Thunder*, 292–93.

32. Whipple, *To the Shores of Tripoli*, 154–57. Preble to Smith, in "Naval Operations against Tripoli," *Barbary*, 4:294–95. Preble Journal, 3 August 1804, *Barbary*, 4:336–38.

33. Tucker, *Dawn Like Thunder*, 302–5. Preble to Smith, in "Naval Operations against Tripoli," *Barbary*, 4:296–97.

34. M. Beaussier to Edward Preble, 6 August 1804, *Barbary*, 4:369–70. Preble to Smith, 7 August, in "Naval Operations against Tripoli," *Barbary*, 4:298–99. Robert Spence to Mrs. Keith Spence, 12 November 1804, *Barbary*, 4:351–52. Tucker, *Stephen Decatur*, 70–71.

35. Journal of John Darby, 7 August 1804, *Barbary*, 4:384–85. Preble Journal, 7 August 1804, *Barbary*, 4:376–77. Whipple, *To the Shores of Tripoli*, 160.

36. Preble to Smith, 9 August, in "Naval Operations against Tripoli," *Barbary*, 4:301. Spencer Tucker claims that Preble rejected the pasha outright (Tucker, *Stephen Decatur*, 73), but his diary clearly describes the counteroffer. Preble Journal, 10 August 1804, *Barbary*, 4:394.

37. Preble Journal, 12 August 1804, 13 August 1804, 14 August 1804, 17 August 1804, *Barbary*, 4:406, 412, 415, 426. Preble to Smith, 24 August 1804, 28 August 1804, in "Naval Operations against Tripoli," *Barbary*, 4:302–3. Whipple, *To the Shores of Tripoli*, 164–65.

38. U.S. Ketch Intrepid Cargo Manifest, 11 August 1804, *Barbary*, 4:398. Edward Preble to Joseph Israel, 22 August 1804, *Barbary*, 4:446. Log of Nathaniel Haraden, 25 August 1804, *Barbary*, 4:461. Preble Journal, 9 July 1804, *Barbary*, 4:260. Log of Nathaniel Haraden, 29 August 1804, *Barbary*, 4:483.

39. Kirsch, *Fireship*. Whipple, *To the Shores of Tripoli*, 165–66. Haraden Log, 30 August 1804, 31 August 1804, *Barbary*, 4:490, 493.

40. Haraden Log, 2 September 1804, *Barbary*, 4:499. Tucker, *Dawn Like Thunder*, 324. "Description by Charles Ridgely," 4 September 1804, *Barbary*, 4:507–9.

41. Preble to Smith, 4 September, in "Naval Operations against Tripoli," *Barbary*, 4:306. Journal of John Darby, 3 September 1804, *Barbary*, 4:506.

42. "Description by Charles Ridgely," 4 September 1804, *Barbary*, 4:507–9. Tucker, *Dawn Like Thunder*, 324–27.

43. Preble to Smith, 4 September, in "Naval Operations against Tripoli," *Barbary*, 4:306. "Description by Charles Ridgely," 4 September 1804, *Barbary*, 4:507–509. Tucker, *Dawn Like Thunder*, 327.

44. Preble to Smith, 4 September, in "Naval Operations against Tripoli," *Barbary*, 4:306.

45. Tucker, *Dawn Like Thunder*, 328–29. Whipple, *To the Shores of Tripoli*, 168–69. The graves, ostensibly of the Americans recovered from the explosion, still exist in Libya today and were the focal point of an effort to return the remains to the U.S. following

the 2011 insurgency that killed Colonel Muammar Gaddafi. Ukman, "U.S. Sailors Buried in Libya Will Stay There, for Now."

46. Preble to Smith, 5 September 1804, in "Naval Operations against Tripoli," *Barbary*, 4:307.

47. Hull Journal, 10 September 1804, *Barbary*, 5:14. Hull Journal, 11 September 1804, *Barbary*, 5:16. Samuel Barron to James Leander Cathcart, 7 September 1804, in *Barbary*, 5:2. Whipple, *To the Shores of Tripoli*, 169–70.

48. Edward Preble to Stephen Decatur, 24 September 1804, *Barbary*, 5:49. Preble to Decatur, 23 December 1804, *Barbary*, 5: 210. Tucker, *Dawn Like Thunder*, 336–37.

49. Tucker, *Stephen Decatur*, 65–72. Nash, *Forgotten Wars*, 280–89. Edward Preble to Robert Smith, *Barbary*, 4:293–308. Treaty of Peace and Amity between the United States and Tripoli, 4 June 1805, *Barbary*, 6:81–82.

50. Whipple and Glenn Tucker both have "birth of . . ." in their subtitles. Recent works that claim the "forgotten status" include Kilmeade and Yaeger, *Thomas Jefferson and the Tripoli Pirates*.

51. Leiner, "Searching for Nelson's Quote," 48–53.

52. Bainbridge to Preble, 1 November 1803, in *Barbary*, 3:171.

Chapter 4. Raiding on the Lakes, 1812–1814

1. Symonds, *The U.S. Navy*, 20–23.

2. Gilje, *Free Trade and Sailors' Rights*, 1–10. Symonds, *The U.S. Navy*, 24.

3. Maloney, "The War of 1812: What Role for Sea Power?," in Hagan, *In Peace and War*, 46–62. Hamilton to Rodgers, 21 May 1812, *W1812*, 1:118–19.

4. Rodgers to Hamilton, 3 June 1812, *W1812* 1:119–122.

5. For recent scholarship and interpretations see Greenert, "Building on a 200-Year Legacy," 32–33; McCranie, *Utmost Gallantry*; Daughan, *1812*; Lambert, *The Challenge*.

6. Long, *"Mad Jack,"* 12–14.

7. Dent to Crowinshield, 31 January 1815, in Alden, *Lawrence Kearny*, 24–26.

8. For examples see Welsh and Skaggs, *War on the Great Lakes*; Skaggs and Altoff, *A Signal Victory*; Skaggs, *Thomas Macdonough*; Hickey, *Don't Give Up the Ship!*; Black, *The War of 1812 in the Age of Napoleon*; Skaggs, *A Signal Victory*; Schroeder, *The Battle of Lake Champlain*.

9. Forbes, *Report of the Trial of Brig. General William Hull*, 126–27. Yanik, *The Fall and Recapture of Detroit*, 115.

10. Merritt, *Journal of Events Principally on the Detroit and Niagara Frontiers*, 14. Armstrong, "Daring Moves on the Niagara," 36–42.

11. Symonds, *Decision at Sea*, 40.

12. Abstract of Chauncey Journal, 7 August 1812 and 24 September 1812, *W1812*, 1:316, 317. Elliot to Hamilton, 9 October 1812, *ASP NA* 1:282–83.

13. Elliot to Hamilton, 9 October 1812, *ASP NA* 1, 282–83. Malcomson, *Warships of the Great Lakes*, 85–86.

14. Merritt, *Journal of Events*, 14. Elliot to Hamilton, 9 October 1812, *ASP NA* 1:282–83.

15. Hickman to Elliot, 8 January 1813, *ASP NA* 1:284.

16. Merritt, *Journal of Events*, 14. Elliot to Hamilton, 9 October 1812, *ASP NA* 1:282–83.

17. Ibid.

18. Hickman to Elliot, 8 January 1813, and Elliot to Chauncey, 10 October 1812, *ASP NA* 1:283–84.

19. Elliot to Hamilton, 9 October 1812, *ASP NA* 1:284.

20. DANFS—*Caledonia*. Skaggs, *Perry*, 105–6, 109.

21. Chauncey to Hamilton, 16 October 1812, *ASP NA* 1:283–84.

22. Maloney, "The War of 1812: What Role for Sea Power?," 59.

23. Malcomson, *Lords of the Lake*, 200–207.

24. Roosevelt identifies this behavior and refers to both the commanders as "timid and dilatory." Roosevelt, *The Naval War of 1812*, pt. 2, 100.

25. Scott, *Memoirs of Lieut.-General Scott*, 113.

26. Chapelle, *The History of the American Sailing Navy*, 249.

27. Malcomson, *Warships of the Great Lakes*, 117–18.

28. Chauncey to Jones, 21 July 1813, RG 45, M125, Roll 30, NARA. Newspaper accounts cited disagree on the armament of *Fox*, report from the *Albany Argus* suggesting an 8-pounder, and *Buffalo Gazette* listing 18. It is reasonable to assume the two schooners were similarly armed, and the 8-pounder is accurate and the 18 a typo or misread note. Chauncey's official correspondence provides no specific detail of the armament of the ships, just their names, likely because they were private vessels and not public warships on a navy mission.

29. *Buffalo Gazette*, 10 August 1813, *W1812* 2:524. British manpower was so low that gunboats during the summer of 1813 were poorly manned by a combination of militia and a handful of regulars and sailors. Edward Baynes, General Order No. 3, 8 June 1813, *DHCNR* vol. 2, pt. 2: 53. Edward Baynes, General Order, 24 July 1813, *DHCNR* vol. 2, pt. 2: 273–74.

30. "Privateering on the St. Lawrence!" *Niles Weekly Register*, 7 August 1813, 374. There are two Grenadier Islands in the region, one on the Canadian side up the St. Lawrence (referenced here) and another near the mouth of the river on the American side (referenced later).

31. Marion Hunter, 18 August 1813, LAC RG 8, Roll 3173, 679: 441. Edward Baynes, General Order, 24 July 1813, LAC RG 8, Roll 3502, 1170: 323.

32. Edward Baynes, General Order, 24 July 1813, LAC RG 8, Roll 3502, 1170: 323. Malcomson, *Lords of the Lake*, 162. *Buffalo Gazette*, 10 August 1813, *W1812*, 2:524.

33. Pearson to Baynes, 22 August 1813, LAC RG 8, Roll 3173, 679: 473. Edward Baynes, General Order, 24 July 1813, LAC RG 8, Roll 3502, 1170: 323. "Privateering on the St. Lawrence!" *Niles Weekly Register*, 7 August 1813.

34. Chauncey to Jones, 21 July 1813, RG 45, M125, Roll 30, NARA. *Buffalo Gazette*, 10 August 1813, *W1812*, 2:524. Malcomson, *Warships of the Great Lakes*, 50, 65.

35. *Buffalo Gazette*, 10 August 1813, *W1812*, 2:524. "Privateering on the St. Lawrence!" *Niles Weekly Register*, 7 August 1813.

36. James Yeo, Proposed Plan for Manning the Gunboats, July 1813, LAC RG 8, Roll 3243, 730: 48–51. Yeo to Prevost, 21 July 1813, LAC RG 8, Roll 3243, 730: 52–54. Edward Baynes, General Order, 24 July 1813, *DHCNR*, vol. 2, pt. 2:273–74. Malcomson, *Warships of the Great Lakes*, 78. For the contentious relationship between Yeo and Prevost see Grodzinski, *Defender of Canada*, 128–30.

37. Roosevelt, *The Naval War of 1812*, pt. 2, 90.

38. Crawford, *W1812*, 3:463, 508.

39. Woolsey to Chauncey, 1 June 1814, *W1812*, 3:511. Creek seen on modern maps as Salmon River.

40. Malcomson, *Lords of the Lake*, 279.

41. Chauncey to Jones, 2 June 1814, RG 45, M125, Roll 37, NARA.

42. Woolsey to Chauncey, 1 June 1814, *W1812*, 3:511.

43. Popham to Yeo, 1 June 1814, *W1812*, 3:509.

44. Popham to Yeo, 1 June 1814, *W1812*, 3:509. Malcomson, *Lords of the Lake*, 280.

45. Woolsey to Chauncey, 1 June 1814, *W1812*, 3:511. Appling to Gaines, 30 May 1814, *W1812*, 3:508–9.

46. Woolsey to Chauncey, 1 June 1814, *W1812*, 3:511.

47. Le Couteur and Graves, *Merry Hearts Make Light Days*, 162. "The Memoirs of James Richardson, RN, 1814–1815," in Malcomson, *Sailors of 1812*, 85.

48. "The Memoirs of James Richardson, RN, 1814–1815," in Malcomson, *Sailors of 1812*,85.

49. Yeo to Drummond, 3 June 1814, *DHCNF*, 4:18. Drummond to Yeo, 6 June 1814, *DHCNF*, 4:19–20. For "smooth" relationship between Yeo and Drummond and aftermath of Sandy Creek see Turner, *British Generals in the War of 1812*, 122, 142.

50. Grodzinski, *Defender of Canada*, 167. Roosevelt, *The Naval War of 1812*, pt. 2, 93–94.

51. DANFS—*Gregory I*.

52. Goodrich, "Our Navy and the West Indian Pirates," 1466.

53. Shaw to Hamilton, 3 February 1812, *W1812*, 1:379.

54. Smith, "A Man with a Country," 156–57.

55. Malcomson, *Lords of the Lake*, 298.

56. Chauncey to Jones, 20 June 1814, *W1812*, 3:528.

57. Chauncey to Jones, 21 July 1813, *W1812*, 2:523–25.

58. Chauncey to Jones, 20 June 1814, *W1812*, 3:529.

59. "A List of His Majesty's Gunboats," James Yeo to Admiralty, April 1814, ADM 1, 2737:78.

60. Initial details in John Hewson to Morrison, 19 June 1814, LAC RG8, Roll 3174, 683: 299. Quote from Drummond to Prevost, 21 June 1814, LAC RG8, Roll 3174, 683: 301.

61. Chauncey to Jones, 20 June 1814, *W1812*, 3:529.

62. Drummond to Prevost, 23 June 1814, LAC RG8, Roll 3174, 683: 303–4.

63. Drummond to Prevost, 23 June 1814, LAC RG8, Roll 3174, 683: 304–5.

64. Chauncey to Jones, 20 June 1814, *W1812*, 3:529.

65. Chauncey to Jones, 7 July 1814, *W1812*, 3:531–32.

66. Wolcott Chauncey to Isaac Chauncey, 18 June 1813, RG 45, M125, 29:82, NARA.

67. Malcomson, *Lords of the Lake*, 120, 122.

68. O'Conner to Freer, 24 November 1813, LAC RG8, Roll 3244, 731: 136–7, NARA. Commodore Chauncey repeatedly refers to the location as "Presque Isle," easily allowing it to be confused with the American base on Lake Erie at Erie, Pennsylvania. This operation occurred at the Canadian village sixty miles west of Kingston. From American reporting, the location of the shipyard is properly identified in *Niles Weekly Register* 6 (16 July 1814): 337.

69. Roosevelt, *The Naval War of 1812*, pt. 2, 94.

70. Chauncey to Jones, 7 July 1814, *W1812*, 3:531.

71. Deposition of Mathias Steele, 31 May 1814, *DHCNR*, 4:16–17. Sinclair to Jones, 13 May 1814, *W18123*, 483–84.

72. Drummond to Prevost, 27 May 1814, *DHCNR*, 4:15–16. Opinion of a Court of Enquiry on the Conduct of Colonel Campbell, 20 June 1814, *DHCNR*, 4:18.

73. Chauncey to Jones, 7 July 1814, *W1812*, 3:532–33.

74. Chauncey to Jones, 7 July 1814, *W1812*, 3:532.

75. Chauncey to Jones, 29 August 1814, *W1812*, 3:595.

76. Gregory to Chauncey, 27 August 1814, *W1812*, 3:596. Daverne to Powell, 28 August 1814, *W1812*, 3:594.

77. Wingfield, Bamford, and Carroll, *Four Years on the Great Lakes*, 119–20. Yeo to Freer, 27 August 1814, LAC RG8, Roll 3244, 733: 59–60.

78. Daverne to Powell, 28 August 1814, *W1812*, 3:594. Wingfield, Bamford, and Carroll, *Four Years on the Great Lakes*, 83–84.

79. Wingfield, Bamford, and Carroll, *Four Years on the Great Lakes*, 120. Gregory to Chauncey, 27 August 1814, *W1812*, 3:596.

80. Daverne to Powell, 28 August 1814, *W1812*, 3:594. Pring to Robinson, 22 October 1814, LAC RG8, Roll 3233, 694: 1w. Carter to Shaw, 7 November 1814, *W1812*, 3:651–52.

81. Smith, "A Man with a Country," 158–60.

82. Malcomson, *Warships of the Great Lakes*, 86, 88. DANFS—*Caledonia*. Gough, *Fighting Sail on Lake Huron and Georgian Bay*, 86–87.

83. Malcomson, *Lords of the Lake*, 282.

84. "Memoirs of James Richardson," in Malcomson, *Sailors of 1812*, 34–37. Chauncey to Jones, 3 July 1813, RG 45, M125, Roll 29, 147, NARA.

85. Dobbs to Yeo, 13 August 1814, *W1812*, 3:588–89.

Chapter 5. Destructive Machines and Partisan Operations

1. Broke to Lawrence, undated 1813, RG 45, M125, Roll 29, 12a, NARA. Philip Broke to Thomas Capel, ADM 1/503, 645–53. (Whether the letter to Capel was actually written by Broke is disputed, since he was severely injured in the battle, see: *W1812*, 2:133). Hickey, *The War of 1812*, 158.

2. Decatur to Jones, undated, RG 45, M125, Roll 29, 3, NARA.

3. Lambert, *The Challenge*, 229–30.

4. Hutcheon, *Robert Fulton*, 41–44. Burgess, *Ships Beneath the Sea*, 36–38.

5. Philip, *Robert Fulton*, 172–81. Hutcheon, *Robert Fulton*, 91–92.

6. Sale, *The Fire of His Genius*, 107–15.

7. Fulton to Madison and to the Members of both Houses of Congress, 26 February 1810, *ASP NA*, 1:211.

8. Fulton, *Torpedo War*.

9. Ibid., 10–11, 13–17.

10. "An Act making an appropriation for the purposes of trying the practical use of the Torpedo or sub-marine explosions," 30 March 1810, *SAL*, 2:569.

11. Hamilton to Fulton, 4 May 1810, RG 45, M209, Roll 4, vol. 10, 192, NARA. Hamilton to Fulton, 14 May 1810, RG 45, M209, Roll 4, vol. 10, 199, NARA.

12. Hamilton to Fulton, 4 September 1810, RG 45, M209, Roll 4, vol. 10, 220, NARA. Hamilton to Varnum, 12 February 1811, *ASP NA*, 1:234–35.

13. Fulton, *Torpedo War*, 50–52.

14. Rodgers Journal, 21 September 1810 and 24 September 1810, *ASP NA*, 1:239–40.

15. Livingston to Hamilton, 7 December 1810, *ASP NA*, 1:237–38. Fulton to Hamilton, 1 February 1811, *ASP NA*, 1:243.

16. Panel to Hamilton, 22 January 1811, *ASP NA*, 1: 235.

17. Fulton to Hamilton, 1 February 1811, *ASP NA*, 1:245.

18. Characterization in Bauer, "John Rodgers: The Stalwart Conservative," in Bradford, *Command under Sail*, 220–47.

19. Hamilton to Fulton, 8 April 1811, RG 45 M209, Roll 4, 10, 322, NARA.

20. Hamilton to Fulton, 13 May 1812, RG 45 M209, Roll 4, 11, 84, NARA.

21. Fulton to Hamilton, 22 June 1812, *W1812*, 1:146–47.

22. Croker to Warren, 20 March 1813, ADM 2/1376, 341–67.

23. Jones to Campbell, Jones to Lewis, Jones to Murray, Jones to Gautier, all 26 February 1813, RG 45, M149, Roll 10, 280–85, NARA. Jones to Dent, 28 February 1813, RG 45 M149, Roll 10, 286, NARA. Jones to Shaw, 1 March 1813, RG45, M149, Roll 10, 287, NARA.

24. "An Act to encourage the destruction of the armed vessels of war of the Enemy," 3 March 1813, *SAL*, 2:816.

25. Ibid.

26. Fulton to Hamilton, 22 June 1812, *W1812*, 1:146–47.

27. Fulton to Madison, 18 March 1813, in Angela Kreider, et al (eds.), *The Papers of James Madison*, Presidential Series, vol. 6, *8 February–24 October 1813* (Charlottesville, VA: University of Virginia Press, 2008), 132–33.

28. Lewis to Hamilton, 22 January 1810, *ASP NA*, 1:239.

29. Lewis to Jones, 28 June 1813, RG 45, M125, Roll 29, 137, NARA.

30. Robert Fulton, "Notes on the Practice of Torpedoes," 26 March 1813, U.S. Naval Academy Museum, Zabriskie Collection, No. 12.

31. Fulton to Jefferson, 7 April 1813, *The Papers of Thomas Jefferson*, 6:55–57.

32. Calderhead, "Naval Innovation in Crisis," 210. Whitehorne, *Battle for Baltimore*, 52–54.

33. Gordon to Jones, 13 March 1813, RG 45 M125, Roll 27, 57, NARA. Neimeyer, *War in the Chesapeake*, 108–9.

34. Latrobe to Fulton, 13 April 1813, Letter book, The Huntington Library, San Marino, Calif., cited in Shomette, "Infernal Machines," 18.

35. Fulton to Jones, 13 April 1813, RG 45, M124, Roll 55, 18, NARA.

36. Robert Fulton to James Monroe, 4 May 1813, USNA Museum, Zabriskie Collection, No. 10.

37. Mix to Madison, 8 April 1813, *Madison Papers*, 6:182–83.

38. Mix to Madison, 15 April 1813, *Madison Papers*, 6:202.

39. Jones to Gordon, 7 May 1813, *W1812*, 2:355.

40. Charles Gordon to William Jones, 10 May 1813, RG 45, M125, Roll 29, 113, NARA.

41. Charles Gordon to William Jones, 19 May 1813, *W1812*, 2:351–52.

42. Charles Gordon to William Jones, 21 June 1813, *W1812*, 2:351–52. Calderhead, "Naval Innovation in Crisis," 217.

43. Gordon to Jones, 6 June 1813, RG 45, M125, Roll 29, 20, NARA.

44. Log of HMS *Victorious*, 5 June 1813, ADM 51/2936. "American Torpedoes," *Liverpool Mercury*, 15 October 1813, No. 120. Laughton, "Talbot, Sir John (c. 1769–1851)," in *Studies in Naval History: Biographies*, 706–7.

45. Cockburn to Warren, 16 June 1813, *W1812*, 2:355. Field, *The Story of the Submarine*, 76. The squadron's concentration near Hampton Roads was confirmed by Gordon when his squadron of schooners returned to Baltimore: Gordon to Jones, 21 June 1813, *W1812*, 2:351–52.

46. Calderhead's "Naval Innovation in Crisis" suggests that Mix did nothing until July 1813 (p. 217). Others have repeated this assumption. But Calderhead makes no use of British sources and misses the reports of *Victorious* and the explanation of Cockburn's retreat to Hampton Roads. The likelihood that Mix used the makeshift torpedo he constructed in Baltimore against *Victorious* also aligns with the fact that the device he built disappears from the records and that he used a Fulton-supplied design in his attack in July.

47. Lewis to Jones, 28 June 1813, RG 45, M124, Roll 56, 107, NARA.
48. Guernsey, *New York City and Vicinity*, 1:279–80.
49. Hardy to Warren, 25 June 1813, ADM 1/504, 51–53. Log of *Ramillies*, 25 June 1813, ADM 53/1126.
50. Ibid.
51. Lewis to Jones, 28 June 1813, RG 45, M124, Roll 56, 107, NARA.
52. Decatur to Jones, 2 July 1813, RG 45, M125, Roll 29, 137, NARA.
53. General Order Number 87, Signed by H. Hotham, 18 July 1813, *W1812*, 2:164.
54. Decatur to Jones, 2 July 1813, RG 45, M125, Roll 29, 137. Shomette, "Infernal Machines," 20.
55. "Boston 13 July," *The Times* (London), 16 August 1813, 4, Issue 8990, Column A.
56. Fulton to Decatur, 5 August 1813, *W1812*, 2:211.
57. Lewis to Jones, 8 August 1813, RG 45, M124, Roll 57, vol. 5, 77, NARA.
58. Mix appointed Sailing Master, June 1813, *W1812*, 2:354.
59. Fulton to Jones, 8 May 1813, RG 45, M124, Roll 55, 82, NARA.
60. *Niles Weekly Register,*vol. 4, 7 August 1813, 365–66. Calderhead, "Naval Innovation in Crisis," 217.
61. "Torpedoes Again!" *Norfolk Herald*, 27 June 1813, reprinted in *Connecticut Courant*, 10 August 1813, 2. Calderhead, "Naval Innovation in Crisis," 217–18. See also Butler, *Defending the Old Dominion*, 298–99.
62. Log of HMS *Plantagenet*, 24 July 1813, ADM 51/2714. "American Torpedoes," *Liverpool Mercury*, Friday, October 15, 1813, Issue 120.
63. *Norfolk Herald*, 27 June 1813.
64. Log of HMS *Plantagenet*, 24 July and 25 July 1813, ADM 51/2714." Torpedoes," *Connecticut Courant*, 14 August 1813, 2.
65. Philip, *Robert Fulton*, 296–97. Fithian, "'For the Common Defense.'"
66. Guernsey, *New York City*, 1:281–82.
67. Hardy to Case, 24 August 1813, Guernsey, *New York City*, 1;285–87. *Connecticut Courant*, 7 Sept 1813, 3.
68. *Connecticut Courant*, 7 Sept 1813, 3. Penny, *The Life and Adventures of Joshua Penny*, 53–55.
69. Case to Hardy, undated, Guernsey, *New York City*, 1:284–85.
70. Hardy to Case, 24 August 1813, Guernsey, *New York City*, 1:285–87. Decatur to Jones, 6 October 1813, RG45, M129, Roll 31, 130, NARA. For discussion of American smuggling and supplying of British blockaders see Joshua A. Smith, "Patterns of Northern New England Smuggling, 1789–1820," in Dudley and Crawford, *The Early Republic and the Sea*, 47–48.
71. Hardy to Terry, 23 August 1813, Guernsey, *New York City*, 1:287. *Connecticut Courant*, 7 September 1813, 3.
72. Fulton to Decatur, 5 August 1813, USNA Museum, Zabriskie Collection, No. 11. Allison, *Stephen Decatur*, 129–30. Design and building of *Demologos* has been covered by a number of historians, notably Chapelle, "Fulton's 'Steam Battery,'" Naval appropriation act of November 1776 in *The History of the American Sailing Navy*.
73. Fulton to Decatur, 4 Sept 1813, USNA Museum, Rosenbach Collection.
74. Fulton to Decatur, 5 August 1813, USNA Museum, Zabriskie Collection, No. 11. Decatur to Fulton, 9 August 1813, *W1812*, 2:212.

75. Fulton and Welden Contract, 30 July 1813, War of 1812 Collection, Lilly Library, Indiana University, Bloomington, Indiana, http://collections.libraries.indiana.edu/warof1812/items/show/1271.
76. Decatur to Fulton, 9 August 1813, *W1812*, 2:212.
77. "New London," 1 Sept 1813, *The War*, vol. 2, no. 12.
78. Roberts, *The British Raid on Essex*, 21–22. Roberts and others make the case that the attack on *La Hogue* is part of what triggered Capel's own irregular raid up the Connecticut River in a small-boat attack on the harbor at Pettipaug.
79. Guernsey, *New York City*, 2:139–40. DeKay, *Battle of Stonington*, 128–31.
80. "Torpedo Boat," *Connecticut Courant*, 5 July 1814, 3.
81. "Loss of Torpedo Boat," *Connecticut Courant*, 5 July 1814, 3. Guernsey, *New York City*, 2:139–40. Shomette, "Infernal Machines," 20–21.
82. Porter to Jones, 7 Sept 1814, RG 45 M125, Roll 39, 27, NARA. O'Neill, "Between the Burning and Bombardment."
83. Whitehorne, *Battle for Baltimore*, 148–50.
84. Fulton Diary, 8–9 Sept 1814, Montagu Collection, New York Public Library, quoted in Hutcheon, *Robert Fulton*, 126.
85. Neimeyer, *War in the Chesapeake*, 217n44.
86. Chauncey to Jones, 19 November 1814, RG 45, M125, Roll 41, 11, NARA.
87. McGowan to Chauncey, 19 November 1814, RG 45, M125, Roll 41, 11, NARA.
88. Ibid.
89. Ibid. Chauncey to Jones, 19 November 1814, RG 45, M125, Roll 41, 11.
90. DeKay, *Battle of Stonington*, 65–66. Lambert, *The Challenge*, 246–47.
91. Sims, "Military Conservatism," in Armstrong, *21st Century Sims*, 104–20.
92. Allison, *Stephen Decatur*, 129.
93. Schroeder, *Commodore John Rodgers*, 81–82.
94. Bauer, "John Rodgers," 220–22, 226–27, 238–39.
95. Guttridge, *Our Country, Right or Wrong*, 33.
96. Schroeder, "Stephen Decatur," 199–219. Decatur to Fulton, 9 August 1813, *W1812*, 2:212.
97. "Jacob Lewis and the New York Flotilla," in *W1812*, 2:39–41. Lewis to Jones, 28 June 1813, RG 45, M124, Roll 56, 106, NARA. Lewis to Jones, 9 August 1813, RG 45, M124, Roll 57, 77, NARA.
98. "Torpedoes Again!" *Norfolk Herald*, 27 June 1813, reprinted in *Connecticut Courant*, 10 August 1813, 2. Fulton, *Torpedo War*, 7–8.
99. Goldenberg, "Blue Lights and Infernal Machines," 385–97. Calderhead, "Naval Innovation in Crisis," 206–21.
100. Benjamin Henry Latrobe to Robert Fulton, 13 April 1813, cited in Shomette, "Infernal Machines," 18. Robert Fulton to William Jones, 13 April 1813, RG 45, M124, Roll 55, 18, NARA. Robert Fulton and James Welden Contract, 28 July 1813, War of 1812 Collection, Lilly Library, Indiana University, Bloomington, Indiana. Philip, *Robert Fulton*, 296.
101. Fulton, "Notes on the Practice of Torpedoes," 26 March 1813, U.S. Naval Academy Museum, Zabriskie Collection, No. 12.
102. Decatur to Fulton, 9 August 1813, *W1812*, 2:212. Lewis to Jones, 28 June 1813, RG 45, M124, Roll 56, 106, NARA. Lewis to Jones, 20 June 1813, RG45, M124, Roll 56, 67,

NARA. Lewis to Jones, 9 August 1813, RG 45, M124, Roll 57, 77, NARA. Lewis to Jones, 6 July 1813, RG45, M124, Roll 56, 140, NARA.

Chapter 6. Pirates and Privateers

1. Symonds, *Navalists and Antinavalists*, 12–13, 196–98. Symonds, *The U.S. Navy*, 31–32.
2. The most comprehensive work on the Second Barbary War is Leiner, *The End of Barbary Terror*. Dey to British consul quoted in Hagan, *This People's Navy*, 92.
3. Hagan, *This People's Navy*, 94. David Long, "The Navy under the Board of Commissioners," in Hagan, *In Peace and War*, 63–64. Symonds, *The U.S. Navy*, 31–32.
4. *Philadelphia Advertiser*, 12 March 1792, *NWIP*, 1:1182.
5. Earle, *Pirate Wars*, 160–80.
6. *NWIP*, 1:1190. Jefferson, Fifth Message to Congress, 3 December 1805, *ASP FR* 1:66.
7. Rodríguez O., *The Independence of Spanish America*, 49–51.
8. The term "junta" has taken on a different and more ominous connotation in modern political parlance; however, at the time it simply meant "committee" or a ruling council that was established locally. Rodríguez O., *Independence of Spanish America*, 51–52.
9. Rodríguez O., *Independence of Spanish America*, 53.
10. Rodríguez O., *Independence of Spanish America*, 110–11, 123–25.
11. Lynch, *The Spanish American Revolutions*, 58–60.
12. McCarthy, *Privateering, Piracy, and British Policy*, 24, 33–34.
13. Cusick, *The Other War of 1812*. McMichael, *Atlantic Loyalties*.
14. Long, *Nothing Too Daring*, 36–56. Allen, *Our Navy and the West Indian Pirates*, 5–16.
15. Mecham, *A Survey of United States–Latin American Relations*, 26–29.
16. Earle, *Pirate Wars*, 212
17. McCarthy, *Privateering, Piracy, and British Policy*, 26, 47.
18. "An Act to Protect the Commerce of the United States and Punish the Crime of Piracy," 3 March 1819, *SAL*, 15th Congress, 510–14.
19. Gilje, *Free Trade and Sailors' Rights*, 299–300.
20. Mecham, *A Survey of United States–Latin American Relations*, 30–31.
21. Head, *Privateers of the Americas*, 64–66. *NWIP*, 3:1479.
22. Thompson to Perry, 20 May 1819, *New American State Papers, Naval Affairs*, 3:5–20.
23. Skaggs, *Perry*, 203–9. Long, *Gold Braid*, 58–59. Allen, *West Indian Pirates*, 18. *NWIP*, 3:1929–30.
24. Soley, "The Autobiography of Commodore Charles Morris," 192–94. Allen, *West Indian Pirates*, 19.
25. A. Welles, T. Perkins, J. Apthorp, J. Hall, N. Goddard, F. Oliver to James Monroe, 1 December 1819, *NWIP*, 3:1935.
26. Goodrich, *NWIP*, 3:1933.
27. "An Act to continue in force 'An act to protect the commerce of the United States and punish the crime of piracy,' and also to make further provisions for punishing the crime of piracy," 15 May 1820, *SAL*, 16th Congress, 600–601.
28. "Statement of Capture of Piratical Vessels" in Smith Thompson to James Monroe, 30 November 1822, *ASP NA*, 1:804.
29. McCarthy, *Privateering, Piracy, and British Policy*, 43.
30. Ibid., 39–41.
31. Long, *Sailor-Diplomat*, 93–96. *NWIP*, 4:91–92.

32. Biddle to Thompson, 6 May 1822, Biddle to Mahy, 30 April 1822, Mahy to Biddle, 2 May 1822, *ASP NA*, 1:8056. Long, *Gold Braid*, 60–62.

33. Allen, *West Indian Pirates*, 27. Statement of Capture of Piratical Vessels, *ASP NA*, 1:804.

34. Sailing vessel with foremast rigged as a brigantine but with a mainmast rigged as a schooner.

35. Gregory to Biddle, 24 August 1822, *ASP NA*, 1:806. Gregory to Thompson, undated, *ASP NA*, 1:806. Long, *Sailor-Diplomat*, 96–100.

36. Bahia Honda, Cuba.

37. Orders from Rowley to Geary, 6 Sept 1822, ADM 1/272, 115.

38. Cassin to Thompson, 10 October 1822, *NWIP*, 6:488–89. Geary to Rowley, 7 October 1822, ADM 1/273, 132. Cassin to Geary, 8 October 1822, ADM 1/273, 132. Allen, *West Indian Pirates*, 35–36.

39. Dale to Thompson, 16 November 1822, *ASP NA*, 1:824. "The Pirates of Cuba, Pirate fight of the 9th inst.," *Niles Weekly Register*, 23:7 December 1822, 211–12.

40. Dale to Thompson, 16 November 1822, *ASP NA*, 1:824.

41. *Niles Weekly Register*, 7 December 1822.

42. Dale to Thompson, 16 November 1822, *ASP NA*, 1:824. *Niles Weekly Register*, 7 December 1822.

43. Dale to Thompson, 16 November 1822, *ASP NA*, 1:824. *Niles Weekly Register*, 7 December 1822.

44. Ibid. Report and newspaper do not specify what states the schooners were from, but Rochester, New York, on Lake Ontario appears to be the most likely homeport for the first, despite its Great Lakes location, and Salem, Massachusetts, the second.

45. *National Advocate* (New York, N.Y.), 10 December 1822. Bradlee, *Piracy in the West Indies*, 31.

46. Monroe to U.S. Senate, *NWIP*, 6:491.

47. Thompson to Pleasants, 11 December 1822, *ASP NA*, 1:822. Schroeder, *Rodgers*, 160–61. Long, *Nothing Too Daring*, 210.

48. Thompson to Pleasants, 11 December 1822, *ASP NA*, 1:822.

49. DANFS—*Fulton I*.

50. Bradlee, *Piracy in the West Indies*, 34.

51. Thompson to Pleasants, 11 December 1822, *ASP NA* 1:822. "Expense of Building Each Vessel Authorized by Act of Jan 2, 1813," Monroe to U.S. Senate, 3 January 1823, *ASP NA*, 1:825–28.

52. Thompson to Porter, 1 February 1823, *NASP*, 5:33–35. Long, *Nothing Too Daring*, 210. Bradlee, *Piracy in the West Indies*, 36. *NWIP*, 7:683.

53. Porter to Thompson, 28 March 1823, *NWIP*, 7:683–84.

54. Porter to de la Torre, 11 March 1823, *ASP NA*, 1:1105–6. Porter to Dionisio Vives, 26 March 1823, *ASP NA*, 1:1107.

55. Translation—Circular Letter from the Captain General of Cuba to Commandants of the several military stations of that island, 10 May 1823, *ASP NA*, 1:1111.

56. Thompson to Porter, 1 February 1823, *NWIP*, 6:494–96.

57. Allen, *West Indian Pirates*, 46–51. Porter to Thompson, 28 March 1823, *NWIP*, 7:683–84. Porter to Thompson, 24 April 1823, *NWIP*, 7:687. Porter to Vives, 13 July 1823, *NWIP*, 7:696.

58. Gray to Rowley, 17 January 1823, ADM 1/273, 60. Canning to Rowley, 7 February 1823, ADM 1/273, 63.

59. Rowley to Admiralty, 29 July 1823, ADM 1/275, 146.

60. Maclean to Owen, 24 Sept 1823, 26 Sept 1823, ADM 1/275, 262.

61. Hunter, *Policing the Seas*, 87–88. Allen, *West Indian Pirates*, 51.

62. Frierson, "The Yellow Fever Vaccine: A History," 77–85.

63. Porter to Southard, 31 August 1823, *ASP NA*, 1:1116. Southard was appointed to replace Smith Thompson when Thompson resigned to become an associate justice, U.S. Supreme Court. Long, *Nothing Too Daring*, 216.

64. Southard to Monroe, 21 Sept 1823, *ASP NA*, 1:1116–17. Orders from Southard to Rodgers, 29 Sept 1823, *ASP NA*, 1:1117.

65. Porter to Chauncey, 27 October 1823, *ASP NA*, 1:1118.

66. Southard to Monroe, 1 December 1823, *ASP NA*, 1:1094.

67. "Message of the President of the United States at the commencement of the first session of the 18th Congress," 2 December 1823, *ASP FR*, 5:248.

68. Long, *Nothing Too Daring*, 217.

69. "Letter from an American in Matanzas," 16 October 1823, *Daily National Intelligencer*, 8 December 1823, No. 3398.

70. Southard to Porter, 2 December 1823, "The Court-Martial of Commodore David Porter," 190–91.

71. *NWIP*, 8:980.

72. Porter to Southard, 14 January 1824, Porter to Southard, 20 January 1824, "The Court-Martial of Commodore David Porter," 217–18.

73. Porter to Southard, 8 April 1824, *ASP NA*, 1:1006.

74. Wilkinson to Porter, 24 April 1824, Lee to Porter, 12 May 1824, *ASP NA*, 1:1006–7.

75. Montgomery to Porter, n.d., *ASP NA*, 1:1009.

76. Porter to Southard 28 May 1824, Porter to Southard, 25 June 1824, "The Court-Martial of Commodore David Porter," 220–21.

77. Southard to Porter, 19 July, 20 July, 28 July, 29 July 1824, "The Court-Martial of Commodore David Porter," 195–96.

78. Porter to Southard, 10 August 1824, *ASP NA*, 1:1009–10.

79. Southard to Porter, 14 October 1824, "The Court-Martial of Commodore David Porter," 197.

80. Southard to Porter, 21 October 1824, "The Court-Martial of Commodore David Porter," 197.

81. Porter to Southard, 15 November 1824, *ASP NA*, 1:1024.

82. Testimony of Charles Platt, 4 May 1925, "The Court-Martial of Commodore David Porter," 15–16.

83. Porter to Southard, 15 November 1824, "The Court-Martial of Commodore David Porter," 225–26.

84. Southard to Porter, 27 December 1824, "The Court-Martial of Commodore David Porter," 197–98.

85. Allen, *West Indian Pirates*, 70–71. Bradlee, *Piracy in the West Indies*, 126–27. Edward Beach, "The Court-Martial of Commodore David Porter," 1401–2.

86. Long, *Nothing Too Daring*, 246–47. "Minutes of the Proceedings," "The Court-Martial of Commodore David Porter," 359–500.

87. Porter to Warrington, 28 January 1825, 29 January 1825, "The Court-Martial of Commodore David Porter," 348–49.

88. Warrington to Southard, 15 February 1825, 10 March 1825, *ASP NA*, 2:103.

89. Sloat to Warrington, 12 March 1825, *ASP NA*, 2,:104.
90. De la Torres to Sloat, undated, *ASP NA*, 2:105–6.
91. Warrington to Southard, 10 March 1825, *ASP NA*, 2:104.
92. Warrington to Southard, 2 April 1825, *ASP NA*, 2:107.
93. Maude to Halstead, 30 March 1825, ADM 1/277, 86. McKeever to Warrington, 1 April 1825, *ASP NA*, 2:107.
94. Warde to Maude, undated, ADM 1/277, 86 encl. McKeever to Warrington, 1 April 1825, *ASP NA*, 2:107–8. Official documents offer no indication whether the landing to avoid the weather had the permission of local authorities. However, if confronted the lieutenants might have easily invoked their status as mariners in danger.
95. Maude to Halstead, 30 March 1825, ADM 1/277, 86. Warrington to Southard, 27 April 1825, *ASP NA*, 2:109.
96. Robert B. Cunningham is the only Cunningham on the navy list of that era with the right seniority to be the officer mentioned in correspondence. He was promoted from midshipman to lieutenant on 13 January 1825 and appears to be the same officer who served with Allen two years prior as a midshipman, or acting lieutenant. Callahan, *List of Officers of the Navy of the United States and of the Marine Corps*, 143.
97. Warde to Maude, undated, ADM 1/277, 86 encl. McKeever to Warrington, 1 April 1825, *ASP NA*, 2:107–108. Account of the attack on *Betsey* and name of sailor in Allen, *West Indian Pirates*, 73–74.
98. Captain's Logs, HMS *Dartmouth*, 28 March 1825, ADM 51/3134. McKeever to Warrington, 1 April 1825, *ASP NA*, 2:107–8.
99. Warrington to Southard, 3 April 1823, *ASP NA*, 2:107.
100. Warrington to Southard, 27 April 1825, *ASP NA*, 2:109. Long, *Gold Braid*, 67.
101. Warrington to Southard, 7 July 1824, *ASP NA*, 2:109. Warrington to Southard, 29 August 1825, in Allen, *West Indian Pirates*, 85–86. For cooperation, see de la Torres to Sloat, 16 March 1825, *ASP NA*, 2:106, and Atkinson to Sloat, 12 March 1825, *ASP NA*, 2:107.
102. Southard to Warrington, 24 May 1825, *ASP NA*, 2:109–10. Jesup to Burch, 13 May 1825, *ASP NA*, 2:110. Bainbridge, Warrington, and Biddle to Southard, 4 November 1825, *ASP NA*, 2:111. Long, *Sailor-Diplomat*, 131.
103. Quincy Adams to Congress, 6 December 1825, *U.S. Congressional Documents and Debates*, 19th Congress, 1st Session, Appendix 2.
104. Allen, *West Indian Pirates*, 87.
105. Long, *Gold Braid*, 151–55, 199–200. Swartz, *U.S.-Greek Naval Relations Begin*.
106. Chapelle, *The History of the American Sailing Navy*, 324, 338, 355.
107. Historical study of the transition from sail to steam almost exclusively begins in the late 1830s and 1840s, usually with a passing reference to *Fulton* at the end of the War of 1812 and no mention of *Sea Gull* or West Indian operations. Hackemer, *The U.S. Navy and the Origins of the Military-Industrial Complex*, 10–20, 45. Tomblin, "From Sail to Steam," 9–14.
108. DANFS—*Sea Gull I*.
109. Chapelle, *The History of the American Sailing Navy*, 367.
110. Swartz, *U.S.-Greek Naval Relations Begin*, 3–6.
111. Wright, "Matthew Perry and the African Squadron," in Barrow, *America Spreads Her Sails*, 80–99.

Chapter 7. First Sumatra Expedition, 1831–1832

1. Alexander Hamilton, "Federalist Paper #11," *The Avalon Project*, Yale Law School. Morison, *The Maritime History of Massachusetts*, 43–45. Gallagher, "Charting a New course for the China Trade," 55–74.
2. Putnam, *Salem Vessels and Their Voyages*, 3–6. Morison, *Massachusetts*, 89–91.
3. Woodbury to Downes, 20 June 1831, *ASP NA*, 3:150.
4. Frederiksen, *American Military Leaders*, 1:230.
5. Journal of Midshipman Henry Wadsworth, 2 June 1803, in *Barbary* 2:435–37.
6. Frederiksen, *American Military Leaders*, 230–31. Long, "David Porter: Pacific Ocean Gadfly," in Bradford, *Command under Sail*, 178–85.
7. DANFS—*Falmouth, Dolphin II, Potomac I*.
8. For discussion of economic opportunity, see Hodge, Cavanaugh, and Nolan, *U.S. Presidents and Foreign Policy*, 69.
9. Schroeder, *Maritime Empire*, 20–24. Hagan, *This People's Navy*, 102.
10. Woodbury to Downes, 20 June 1831, *ASP NA*, 3:150.
11. Ibid.
12. Silsbee, Peckman, and Stone to Jackson, 20 July 1831, RG 45, M124, Roll 79, 94, NARA.
13. Woodbury to Silsbee, Peckman, and Stone, 23 July 1831, *ASP NA*, 3:152.
14. Silsbee to Woodbury, 27 July 1831, RG 45, M124, Roll 79, 106, NARA.
15. Endicott, "Narrative of the Piracy, and Plunder of the Ship Friendship," 17.
16. Reynolds, *Voyage of the United States Frigate Potomac*, 90.
17. Statement of Captain Endicott, *ASP NA*, 3:155.
18. Excerpt of Friendship Log Book, 7 February 1831, *ASP NA*, 3:154.
19. Endicott, "Narrative," 18–20.
20. Excerpt of Friendship Logbook, *ASP NA*, 3:154. Endicott, "Narrative," 24–25. Silsbee, Peckman, and Stone to Jackson, 20 July 1831, RG 45, M124, Roll 79, 94, NARA.
21. Woodbury to Downes, 27 June 1831. in *ASP NA*, 3:150–51.
22. Silsbee, Peckman, and Stone to Jackson, 20 July 1831, RG 45, M124, Roll 79, 94, NARA. *Putnam* events detailed in Putnam, *Salem Vessels and Their Voyages*, 23–25. *Marquis de Somerulas* events detailed in Phillips, *Pepper and Pirates*, 55–60.
23. Preble, *The Cruise of the United States Frigate Essex*.
24. Paullin, "Early Voyages of American Naval Vessels to the Orient—I," 442–63. Stewart, *A Visit to the South Seas in the U.S. Ship Vincennes*.
25. Woodbury to Downes, 27 July 1831, 31 July 1831, 3 August 1831, 6 August 1831, and 8 August 1831, RG 45, M149, Roll 20, 52, 53, 54, 55, NARA.
26. Woodbury to Downes, 9 August 1831, *ASP NA*, 3:153.
27. Warriner, *Cruise of the United States Frigate Potomac*, 12. Woodbury to Downes, 6 August 1831, RG 45, M149, Roll 20, 54, NARA.
28. Warriner, *Cruise of the United States Frigate Potomac*, 12, 14, 16–17, 19.
29. *Lexington* was launched in 1825, and earlier in the summer of 1831 she was deployed to the South Atlantic Squadron, which commonly operated out of Rio de Janeiro. DANFS—*Lexington II*. Reynolds, *Voyage of the United States Frigate Potomac*, 32–33. John Downes to Levi Woodbury, 26 October 1832, RG 45, M125, Roll 175, NARA.
30. Reynolds, *Voyage of the United States Frigate Potomac*, 59.
31. Ibid., 61–63.
32. Warriner, *Cruise of the United States Frigate Potomac*, 43. Reynolds, *Voyage of the United States Frigate Potomac*, 95–96.

33. "Character and condition of the population and country at Quallah Batu, in the Island of Sumatra," *ASP NA*, 3:155–56.

34. Downes to Woodbury, 13 February 1833, RG 45, M125, Roll 156, 33, NARA.

35. Reynolds, *Voyage of the United States Frigate Potomac*, 93. Reynolds served as Downes's clerk during the second half of *Potomac*'s service on Pacific Station. His observations corroborate the letters and reports Downes sent to Washington, but this could easily be because he helped draft the letters and reports. The story related to Downes and his officers in Cape Town was likely that of a British East India Company expedition to Muckie in 1804. An initial landing party was attacked by local defenders in June, and a larger expedition under Sir John Shaw returned to Muckie and seized the forts and burned the town in retaliation. Shaw to Marine Board, 3 August 1804, in Stocqueler, *Memoirs and Correspondence of Major-General Sir William Nott*, 1:13.

36. Warriner, *Cruise of the United States Frigate Potomac*, 65.

37. Initial plans related in Reynolds, *Voyage of the United States Frigate Potomac*, 97–98; and Warriner, *Cruise of the United States Frigate Potomac*, 66–67.

38. Warriner, *Cruise of the United States Frigate Potomac*, 65.

39. Ibid., 65, 67.

40. Medical report from J. M. Foltz, "Medical Statistics of the Crew of the Frigate Potomac, during a voyage round the world," appendix in Reynolds, *Voyage of the United States Frigate Potomac*, 539.

41. Downes to Woodbury, 17 February 1832, *ASP NA*, 3:156.

42. Details of changes to *Potomac* in Reynolds, *Voyage of the United States Frigate Potomac*, 104; and Warriner, *Cruise of the United States Frigate Potomac*, 71.

43. Midshipman James L. Parker, 3 February 1832, *Journal of a Cruise Around the World*. Reynolds, *Voyage of the United States Frigate Potomac*, 103. A proa is a coastal watercraft of the Indo-Pacific region, commonly rigged with outriggers and sailed by "shunting" instead of "tacking." For a roughly contemporary nineteenth-century description see Folkard, *The Sailing Boat*, 260–63.

44. Parker, *Journal*, 4 February 1832. Warriner, *Cruise of the United States Frigate Potomac*, 71–72.

45. Downes to Woodbury, 17 February 1832, *ASP NA*, 3:156.

46. Warriner, *Cruise of the United States Frigate Potomac*, 77.

47. Parker, *Journal*, 6 February 1832. Downes to Woodbury, 17 February 1832, *ASP NA*, 3, 156. Reynolds, *Voyage of the United States Frigate Potomac*, 105–6. Warriner, *Cruise of the United States Frigate Potomac*, 78–79. The Parker Journal appears to be off by a day on the dates of entries, as Reynolds, Warriner, and correspondence from Downes to Woodbury all agree on dates.

48. Downes to Woodbury, 13 February 1833, RG 45, M125, Roll 156, 33, NARA.

49. Downes to Woodbury, 17 February 1832, *ASP NA*, 3:156. Reynolds, *Voyage of the United States Frigate Potomac*, 106.

50. Warriner, *Cruise of the United States Frigate Potomac*, 81.

51. Reynolds, *Voyage of the United States Frigate Potomac*, 110. Map of village in Warriner, *Cruise of the United States Frigate Potomac*, 82–83.

52. Parker, *Journal*, 6 February 1832. Warriner, *Cruise of the United States Frigate Potomac*, 82.

53. Shubrick to Downes, 6 February 1832, *ASP NA*, 3:157. Warriner, *Cruise of the United States Frigate Potomac*, 83. Reynolds, *Voyage of the United States Frigate Potomac*, 108–9.

54. Shubrick to Downes, 6 February 1832, *ASP NA*, 3:157. Parker, *Journal*, 6 February 1832. Reynolds, *Voyage of the United States Frigate Potomac*, 109.

55. Warriner, *Cruise of the United States Frigate Potomac*, 83–84. Shubrick to Downes, 6 February 1832, *ASP NA*, 3:157.

56. Reynolds, *Voyage of the United States Frigate Potomac*, 111–12. Shubrick to Downes, 6 February 1832, *ASP NA*, 3:157.

57. Shubrick to Downes, 6 February 1832, *ASP NA*, 3:157. Warriner, *Cruise of the United States Frigate Potomac*, 88. According to Warriner this was the location of the first confirmed killing of Acehnese women, though from his account it appeared they were active combatants.

58. Reynolds, *Voyage of the United States Frigate Potomac*, 112. Shubrick to Downes, 6 February 1832, *ASP NA*, 3:157.

59. Shubrick to Downes, 6 February 1832, *ASP NA*, 3:157.

60. Warriner, *Cruise of the United States Frigate Potomac*, 90. Reynolds, *Voyage of the United States Frigate Potomac*, 112–13.

61. Warriner, *Cruise of the United States Frigate Potomac*, 90.

62. Reynolds, *Voyage of the United States Frigate Potomac*, 113. Shubrick to Downes, 6 February 1832, *ASP NA*, 3:157.

63. Parker, *Journal*, 6 February 1832. Reynolds, *Voyage of the United States Frigate Potomac*, 113–14.

64. Shubrick to Downes, 6 February 1832, *ASP NA*, 3, 157.

65. Foltz, "Medical Statistics," in Reynolds, *Voyage of the United States Frigate Potomac*, 539.

66. Warriner, *Cruise of the United States Frigate Potomac*, 106–9.

67. Reynolds, *Voyage of the United States Frigate Potomac*, 121–22.

68. Parker, *Journal*, 8 February 1832. Warriner, 110. Reynolds, *Voyage of the United States Frigate Potomac*, 124. Parker continues to have his dates off by a day in his journal.

69. Downes to Woodbury, 17 February 1832, *ASP NA*, 3:156.

70. Reynolds, *Voyage of the United States Frigate Potomac*, 125–26. Warriner, *Cruise of the United States Frigate Potomac*, 111.

71. Parker, *Journal*, 14 February 1832. Reynolds, *Voyage of the United States Frigate Potomac*, 130.

72. Reynolds, *Voyage of the United States Frigate Potomac*, 130. Warriner, *Cruise of the United States Frigate Potomac*, 112, 114–15.

73. Downes to Woodbury, 17 February 1832, *ASP NA*, 3:156.

74. Warriner, *Cruise of the United States Frigate Potomac*, 129–31.

75. Parker, *Journal*, 18 February 1832. Warriner, *Cruise of the United States Frigate Potomac*, 125, 127.

76. *New York Evening Post*, 5 July 1832, 2, col. 5. Discussion of Washington's political response in Belohlavek, "Let the Eagle Soar!," 41–42.

77. Meacham, *American Lion*, 215.

78. Andrew Jackson, Fourth Annual Message, 4 December 1832, *Messages of General Andrew Jackson*, 176. Woodbury to Downes, 16 July 1832, RG 45, M149, Roll 21, 59, NARA.

79. Downes to Woodbury, 13 February 1833, RG 45, M 125, Roll 156, 33, NARA. Downes to Woodbury, 17 February 1832, *ASP NA*, 3:156.

80. Ibid.

81. Only one scholarly article examines the events at Kuala Batu, David F. Long's "Martial Thunder: The First Official American Intervention in Asia" in *Pacific Historical Review*. Written at the end of the Vietnam War, the article compares 1970s American involvement in Southeast Asia and *Potomac*'s mission to Sumatra. Through that lens, Long describes Downes's explanation of his actions as "lame excuses." This analysis may have appeared viable through the lens of Washington, D.C., politics in the 1970s, but Long's article makes no attempt to look at the operational challenges faced by Downes, or the full historical record of the voyage to Kuala Batu and events on the ground. Unfortunately, other historians have taken Long's examination and incorporated his conclusions in their work, including Schroeder in *Shaping a Maritime Empire*, and Jampoler in *Embassy to the Eastern Courts*. However, despite Long's attempt to make Downes into a nineteenth-century William Westmoreland, when viewed operationally Downes's explanations hold more water.

82. Downes to Woodbury, 26 October 1832, RG 45, M125, Roll 175, NARA. Parker, *Journal*, 18 February 1832.

83. Reynolds, *Voyage of the United States Frigate Potomac*, 122–23.

84. Long, *Nothing Too Daring*, 121–22.

85. Pratt, *Preble's Boys*, 219.

86. Farragut, *The Life of David Glasgow Farragut*, 31.

87. Porter, *Journal of a Cruise*, 358. "Character and condition . . . ," *ASP NA*, 3:155–56

88. Shubrick to Downes, 6 February 1832, *ASP NA*, 3:157.

89. Discussion of warrant officers in McKee, *A Gentlemanly and Honorable Profession*, 33–34.

90. Putnam, *Salem Vessels and Their Voyages*, 110–24.

91. Reynolds, *Voyage of the United States Frigate Potomac*, 224–25.

92. Putnam, *Salem Vessels and Their Voyages*, 111–23.

93. Phillips, *Pepper and Pirates*, 104–5.

Chapter 8. Return to Sumatra

1. Van Iseghen to Revely, 12 October 1838, *Penang Gazette*, in Taylor, *A Voyage Round the World*, 255–56. "Narrative of George Whitmarsh," republished in Putnam, *Salem Vessels and Their Voyages*, 123–24.

2. Phillips, *Pepper and Pirates*, 108–9.

3. "Narrative of George Whitmarsh," in Putnam, *Salem Vessels and Their Voyages*, 123–24.

4. Iseghen to Revely, 12 October 1838, 255–56.

5. "Narrative of George Whitmarsh," 123–24.

6. Taylor, *A Voyage Round the World*, 10, 243. Read to Paulding, 1 November 1828, RG 45, M125, Roll 249, 62, NARA.

7. DANFS—*Columbia I, John Adams I*.

8. Paullin, "Early Voyages of American Naval Vessels to the Orient IX: The Cruise of Commodore Read," 1074.

9. Taylor, *A Voyage Round the World*, 253–57. An Officer of the U.S. Navy, *Around the World*, 2:43. Colombo and Prince of Wales Island were British, but were colonies of the British East India Company in 1838 rather than Crown colonies. Prince of Wales Island was referred to by its local name, Penang, until 1792.

10. Taylor, *A Voyage Round the World*, 260. Murrell, *Cruise of the Frigate Columbia*, 91–92. Read to James Paulding, 2 December 1838, RG 45, M125, Roll 249, 63, NARA.

11. Taylor, *A Voyage Round the World*, 261. Baker's service in the Second Seminole War in Aldrich, *History of the United States Marine Corps*, 72, 75. Murrell, *Cruise of the Frigate Columbia*, 92–93 (emphasis in original).

12. Taylor, *A Voyage Round the World*, 262–63.

13. Read to Paulding, 12 January 1839, RG 45, M125, Roll 249, 49, NARA. Taylor, *A Voyage Round the World*, 323.

14. Taylor, *A Voyage Round the World*, 326–27.

15. Read to Paulding, 12 January 1839, RG 45, M125, Roll 249, 49, NARA. Taylor, *A Voyage Round the World*, 328.

16. Read to Paulding, 12 January 1839, RG 45, M125, Roll 249, 49, NARA. Taylor, *A Voyage Round the World*, 329.

17. Taylor, *A Voyage Round the World*, 329.

18. Ibid., 330

19. Read to Paulding, 12 January 1839, RG 45, M125, Roll 249, 49, NARA.

20. Taylor, *A Voyage Round the World*, 331.

21. Read to Paulding, 12 January 1839, RG 45, M125, Roll 249, 49, NARA. Read to Wyman, 23 December 1838, reprinted in Taylor, *A Voyage Round the World*, 333.

22. Taylor, *A Voyage Round the World*, 336.

23. Read to Paulding, 12 January 1838, RG 45, M125, Roll 249, 49, NARA. Taylor, *A Voyage Round the World*, 338.

24. Taylor, *A Voyage Round the World*, 340.

25. An Officer, *Around the World*, 2:55.

26. Taylor, *A Voyage Round the World*, 340–41.

27. An Officer, *Around the World*, 2:56.

28. Read to Paulding, 12 January 1839, RG 45, M125, Roll 249, 49, NARA. Taylor, *A Voyage Round the World*, 345.

29. Taylor, *A Voyage Round the World*, 345. An Officer, *Around the World*, 2:56–57.

30. Taylor, *A Voyage Round the World*, 346–48. An Officer, *Around the World*, 2:58–59. Taylor, *A Voyage Round the World*, 349.

31. Read to Paulding, 12 January 1839, RG 45, M125, Roll 249, 49, NARA.

32. An Officer, *Around the World*, 2:59. Taylor, *A Voyage Round the World*, 349–50.

33. Read to Paulding, 12 January 1839, RG 45, M125, Roll 249, 49, NARA. Taylor, *A Voyage Round the World*, 350.

34. Taylor, *A Voyage Round the World*, 350–51. An Officer, *Around the World*, 2:60. Read to Paulding, 12 January 1839, RG 45, M125, Roll 249, 49, NARA.

35. An Officer, *Around the World*, 2:60. Read to Paulding, 12 January 1839, RG 45, M125, Roll 249, 49, NARA. The correspondent in *Around the World* reports that the ships were about two hundred rods from the shore and Read states in his report that they were a cable length, both approximately two hundred yards distant.

36. Read to Paulding, 12 January 1839, RG 45, M125, Roll 249, 49, NARA. Taylor, *A Voyage Round the World*, 356. Murrell, *Cruise of the Frigate Columbia*, 108.

37. Taylor, *A Voyage Round the World*, 358.
38. Wyman to Read, 1 January 1839, RG 45, M125, Roll 249, 49 [enclosure], NARA. Murrell, *Cruise of the Frigate Columbia*, 108. Read to Paulding, 12 January 1839, RG 45, M125, Roll 249, 49, NARA.
39. Wyman to Read, 1 January 1839, RG 45, M125, Roll 249, 49 [enclosure], NARA. Murrell, *Cruise of the Frigate Columbia*, 110–11. An Officer, *Around the World*, 2:61.
40. An Officer, *Around the World*, 2:61–62.
41. Ibid., 2:63.
42. Murrell, *Cruise of the Frigate Columbia*, 111. An Officer, *Around the World*, 2:64.
43. An Officer, *Around the World*, 2:65.
44. Wyman to Read, 1 January 1839, RG 45, M125, Roll 249, 49 [enclosure], NARA. Knowledge of British and Dutch fate in: An Officer, *Around the World*, 2:66.
45. Read to Paulding, 12 January 1839, RG 45, M125, Roll 249, 49, NARA. Taylor, *A Voyage Round the World*, 362.
46. Wyman to Read, 1 January 1839, RG 45, M125, Roll 249, 49 [enclosure], NARA.
47. Read to Paulding, 12 January 1839, RG 45, M125, Roll 249, 49, NARA. Taylor, *A Voyage Round the World*, 365. An Officer, *Around the World*, 2:66.
48. Taylor, *A Voyage Round the World*, 366 (error in original published text marks the page as 466).
49. Ibid., 368.
50. Read to Paulding, 12 January 1839, RG 45, M125, Roll 249, 49, NARA.
51. Taylor, *A Voyage Round the World*, 368.
52. An Officer, *Around the World*, 2:68. Taylor, *A Voyage Round the World*, 366.
53. Taylor, *A Voyage Round the World*, 367. An Officer, *Around the World*, 2:68.
54. Read to Paulding, 12 January 1839, RG 45, M125, Roll 249, 49, NARA.
55. Read to Po Chute Abdullah, 5 January 1839, RG 45, M125, Roll 249, 49 [enclosure], NARA. Taylor, *A Voyage Round the World*, 366–67.
56. Read to Paulding, 12 January 1839, RG 45, M125, Roll 249, 49, NARA.
57. Ibid.
58. Taylor, *A Voyage Round the World*, 369.
59. An Officer, *Around the World*, 2:71.
60. Ibid., 2:69–70.
61. Taylor, *A Voyage Round the World*, 370–71.
62. Datu Buggah, Datu Buggenah, Datu Modah, Datu Umpate to Read, 8 January 1839, RG 45, M125, Roll 249, 49 [enclosure], NARA.
63. Taylor, *A Voyage Round the World*, 371.
64. Read to Paulding, 12 January 1839, RG 45, M125, Roll 249, 49, NARA. Taylor, *A Voyage Round the World*, 375.
65. An Officer, *Around the World*, 2:73–74, 75. Taylor, *A Voyage Round the World*, 375–76. Read to Paulding, 12 January 1839, RG 45, M125, Roll 249, 49, NARA.
66. An Officer, *Around the World*, 2:75–77. Taylor, *A Voyage Round the World*, 376–81.
67. Taylor, *A Voyage Round the World*, 382–84.
68. Po Quallah to Read, 11 January 1839, RG 45, M125, Roll 249, 49 [enclosure], NARA. The mathematics professor on board *Columbia*, J. Henshaw Belcher, was present at the signing as a witness. The near-mirror descriptions of the event in Taylor's memoir and the unattributed *Around the World, a Narrative of a Voyage in the East India Squadron, under Commodore George C. Read*, suggests he was the author.

69. Taylor, *A Voyage Round the World*, 385–86. An Officer, *Around the World*, 2:77–78.
70. An Officer, *Around the World*, 2:68–69. Read to Paulding, 12 January 1839, RG 45, M125, Roll 249, 49, NARA.
71. An Officer, *Around the World*, 2:344.
72. Summary of American pepper trade and vessel sailings to the 1850s in Putnam, *Salem Vessels and Their Voyages*, 126–36.
73. Woodbury to Downes, 20 June 1831, *ASP NA*, 3:150–52.
74. Orders from Dickerson to Read, 13 April 1838, in Paullin, "Early Voyages . . . IX," 1074. Dickerson left office as secretary, turning over responsibilities to James Paulding, a month after Read and his squadron set sail.
75. Woodbury to Downes, 9 August 1831, *ASP NA*, 3:153.
76. "Character and condition of the population and country at Quallah Batoo, in the Island of Sumatra," *ASP NA*, 3:155–56. Downes to Woodbury, 13 February 1833, RG 45, M 125, Roll 156, 33, NARA. Reynolds, *Voyage of the United States Frigate Potomac*, 93. Warriner, *Cruise of the United States Frigate Potomac*, 65.
77. *New York Evening Post*, July 5, 1832, 2, col. 5. Discussion of Washington's political response in Belohlavek, "'Let the Eagle Soar!'," 41–42. Andrew Jackson, Fourth Annual Message, 4 December 1832, *Messages of General Andrew Jackson*, 176. Woodbury to Downes, 16 July 1832, RG 45, M149, Roll 21, 59, NARA.
78. An Officer, *Around the World*, 2:49. Taylor, *A Voyage Round the World*, 320, 322–23. Read to Paulding, 12 January 1839, RG 45, M125, Roll 249, 49, NARA. Read to Wyman, 20 December 1832, quoted in Taylor, *A Voyage Round the World*, 323.
79. Read to Paulding, 12 January 1839, RG 45, M125, Roll 249, 49, NARA. Datu Buggah, Datu Buggenah, Datu Modah, Datu Umpate to Read, 8 January 1839, RG 45, M125, Roll 249, 49 [enclosure], NARA. Read to Po Chute Abdullah, 5 January 1839. Po Quallah to Read, 11 January 1839, RG 45, M125, Roll 249, 49 [enclosure], NARA. Paullin, "Early Voyages . . . IX," 1081. For discussion of legality of non-"treaty" documents see Long, *Gold Braid*, 3–4.
80. Putnam, *Salem Vessels and Their Voyages*, 124–38.
81. Kearney to Upshur, 4 October 1841, 25 January 1842, *NASP*, 3:370, 371–74.
82. McKee, *Edward Preble*, 176, 237–38, 311–12. Long, *Nothing Too Daring*, 203–29. Yoshihara and Holmes, *Red Star Over the Pacific*, 9–10.
83. Long, *Gold Braid*, 218, 363–64.
84. Philbrick, *Sea of Glory*, 201–30. Long, *Gold Braid*, 300–305.
85. Bauer, *Surfboats and Horse Marines*, 171–83. Dupont, "The War with Mexico," 419–37.

Conclusion

1. Beeler, "The State of Naval History," 13.
2. McKee, *A Gentlemanly and Honorable Profession*, 47.
3. Ibid., 155–56, 215. Karsten, *Naval Aristocracy*, 3–5, 23–25.
4. Till, *Seapower*, 30–41. For parallel examples from British history see Beeler, "Maritime Policing and the Pax Britannica," 1–20, and Lambert, "The Limits of Naval Power," 173–90.
5. Buker, *Swamp Sailors in the Second Seminole War*. Swartz, *U.S.-Greek Naval Relations Begin*. Long, *Gold Braid*, 303–5. Johnson, *Far China Station*, 76–94. Long, "Mad Jack."

6. Healy, *Gunboat Diplomacy in the Wilson Era*. Cole, *Gunboats and Marines*. Sherwood, *War in the Shallows*.
7. See also Hattendorf, "The United States Navy in the Twenty-First Century," 285–97.
8. Crowl, "Alfred Thayer Mahan: The Naval Historian" in Paret, *Makers of Modern Strategy*, 8–9.
9. Colomb, *Naval Warfare*, 254–56n6.
10. Ibid., 22–25, 251–52.
11. Corbett, *Some Principles of Maritime Strategy*, 280–81.
12. Demonstrated by late twentieth-century assessments of Colomb that attempt to revive his work but focus entirely on what he had to say about the blue-water and fleet actions, ignoring over half of the content of his book. See Gough, "The Influence of Sea Power Upon History Revisited," 55–63.
13. Mahan, *Influence of Sea Power upon History*, 31–32, 539–40. Hagan, *This People's Navy*, xi–xii. Baer, *One Hundred Years of Sea Power*, 12–18. Till, *Seapower*, 73–76.
14. Bradford, "Guerre de Razzia," 2.
15. Grenier, *The First Way of War*.
16. Bradford, "Guerre de Razzia," 2.
17. Ibid., 1–3.
18. McKee, *A Gentlemanly and Honorable Profession*, 210–15.
19. Goldrick, "The Problems of Modern Naval History," in Hattendorf, *Doing Naval History*, 22.
20. Ibid., 23.
21. Marder, *From the Dardanelles to Oran*, 33–63.
22. Calvert, *Naval Profession*, 57–60. Haynes, *Toward a New Maritime Strategy*, 245.
23. Elias, *The Genesis of the Naval Profession*, 115–20. Haynes, *Toward a New Maritime Strategy*, 243.
24. Karsten, *Naval Aristocracy*, 267–68. Calvert, *Naval Profession*, 129–31.
25. Recent examples that examine the Age of Sail era include Daughan, *If by Sea*; Toll, *Six Frigates*; Dull, *American Naval History*.
26. Examples include Hagan, *American Gunboat Diplomacy and the Old Navy*; Schroeder, *Shaping A Maritime Empire*; Long, *Gold Braid and Foreign Relations*.

Bibliography

Archival Sources

Great Britain

The National Archives, Kew, England
  Admiralty, and Ministry of Defence, Navy Dept: Correspondence and Papers (ADM 1)
  Admiralty: Out-Letters (ADM 2)
  Admiralty: Captains' Logs (ADM 51)
  Admiralty, and Ministry of Defence, Navy Department: Ships' Logs (ADM 53)

United States

Library of Congress, Manuscript Division, Washington, D.C.
  Edward Preble Papers (MSS 36814)
  Porter Family Papers (MSS 53041)
National Archives and Records Administration, Washington, D.C.
  Letters Received by the Secretary of the Navy: Captains Letters, 1805–1861, 1866–1885.
    Record Group 45, Microfilm Series M125.
  Letters Received by the Secretary of the Navy: Miscellaneous Letters, 1805–1861,
    1866–1884. Record Group 45, Microfilm Series M124.
  Letters Sent by the Secretary of the Navy to Officers, 1798–1868, Record Group 45,
    Microfilm Series M149.
  Miscellaneous Letters Sent by the Secretary of the Navy, 1798–1886, Microfilm Series
    M209.
U.S. Naval Academy Museum, Annapolis, Md.
  Papers of Robert Fulton (1765–1815)
      Zabriskie Collection (45.70.31)
      Rosenbach Collection (42.6.14)

Canada

Library and Archives of Canada, Ottawa, Ontario
  British Military and Naval Records, Record Group 8, C Series (RG 8, C Series)

Primary Sources

An Officer of the U. S. Navy. *Around the World, a Narrative of a Voyage in the East India Squadron, under Commodore George C. Read.* New York: C. S. Francis, 1840.

Bauer, K. Jack, ed.. *The New American State Papers, Naval Affairs.* 10 vols. Wilmington, Del.: Scholarly Resources, 1981.

Bradford, James, ed., *The Papers of John Paul Jones.* Microfilm Edition. Cambridge, UK: Chadwyck-Healey, 1986.

Callahan, Edward William. *List of Officers of the Navy of the United States and of the Marine Corps from 1775 to 1900: Comprising a Complete Register of All Present and Former Commissioned, Warranted, and Appointed Officers of the United States Navy and of the Marine Corps, Regular and Volunteer.* New York: Haskell House, 1969.

Clark, William Bell, William James Morgan, and Michael J. Crawford, eds. *Naval Documents of the American Revolution, Vols. 1–12.* Washington, D.C.: Naval History Division, Dept. of the Navy, 1964–2015.

Crackel, Theodore J., et al., eds. *The Papers of George Washington Digital Edition.* Charlottesville: University of Virginia Press, 2007–2018.

Cruikshank, E. A., ed. *The Documentary History of the Campaign on the Niagara Frontier.* 4 vols. Lundy's Lane Historical Society. New York: Arno Press, 1971.

Deasy, Richard M. *William Staples' The Documentary History of the Destruction of the Gaspee, Introduced and Supplemented by Richard M. Deasy.* Providence: Rhode Island Publications Society, 1990.

De Selding, Charles. *Documents, Official and Unofficial, Relating to the Case of the Capture and Destruction of the Frigate Philadelphia, at Tripoli, on the 16th February, 1804.* Washington, D.C.: J. T. Towers, 1850.

Dudley, William S., Michael J. Crawford, Christine F. Hughes, Tamara Moser Melia, Charles Brodine, Carolyn M. Stalling, eds. *The Naval War of 1812: A Documentary History.* 3 vols. Washington, D.C.: Naval Historical Center, Dept. of Navy, 1985–2002.

Endicott, Charles. "Narrative of the Piracy, and Plunder of the Ship Friendship, of Salem, on the West Coast of Sumatra, in February 1831, and the Massacre of Part of Her Crew: Also Her Re-Capture out of the Hands of the Malay Pirates." *Historical Collections of the Essex Institute* 1, no. 1 (April 1859): 15–32.

Examination Papers: United States Naval Academy, 1881–1882. Washington, D.C.: U.S. Government Printing Office, 1883.

Forbes, James G., ed. *Report of the Trial of Brig. General William Hull Commanding the North-Western Army of the United States.* New York: Eastburn Kirk, 1900.

Goodrich, Caspar. "Our Navy and the West Indian Pirates: A Documentary History." 12 parts. U.S. Naval Institute *Proceedings* 42, nos. 4 to 43 (July 1916–September 1917).

Green, Ezra. *Diary of Ezra Green, M.D., Surgeon on Board the Continental Ship-of-War Ranger," under Capt. John Paul Jones, from November 1, 1777, to September 27, 1778.* Boston: D. Clapp & Son, 1875.

Guernsey, R. S. *New York City and Vicinity during the War of 1812–15, Being a Military, Civic and Financial Local History of That Period.* 2 vols. New York: C. L. Woodward, 1889.

Hopkins, Esek, and Alverda Beck, ed. *The Letter Book of Esek Hopkins: Commander-in-Chief of the United States Navy, 1775–1777.* Providence, R.I.: Rhode Island Historical Society, 1932.

Jackson, Andrew. *Messages of Gen. Andrew Jackson, with a Short Sketch of His Life.* Boston: Otis Broaders, 1837.

Jefferson, Thomas. *The Papers of Thomas Jefferson, Retirement Series*. Edited by J. Jefferson Looney. Princeton, N.J.: Princeton University Press, 2006.

Jones, John Paul, and Gerard W. Gawalt, John R. Sellers, eds. *Memoir of the American Revolution Presented to King Louis XVI of France*. Washington, D.C.: American Revolution Bicentennial Office, Library of Congress, 1979.

Knox, Dudley, ed. *Naval Documents Related to the Quasi-War with France: Naval Operations, February 1797–December 1801*. 7 vols. Washington, D.C.: Office of Naval Records and Library, 1935–38.

———. *Naval Documents Related to the United States Wars with the Barbary Powers, Naval Operations Including Diplomatic Background*. 6 vols. Washington, D.C.: Office of Naval Records and Library, 1939–44.

Le Couteur, John, and Donald E. Graves, ed. *Merry Hearts Make Light Days: The War of 1812 Journal of Lieutenant John Le Couteur, 104th Foot*. Ottawa, Ont.: Carleton University Press, 1994.

Malcomson, Robert, ed. *Sailors of 1812: Memoirs and Letters of Naval Officers on Lake Ontario*. Youngstown, N.Y.: Old Fort Niagara Association, 1997.

Merritt, William Hamilton, ed. *Journal of Events Principally on the Detroit and Niagara Frontiers, during the War of 1812*. St. Catharine's, Canada West: Historical Society, British North America, 1863.

Miller, David Hunter, ed. *Treaties and Other International Acts of the United States of America*. 8 vols. Washington, D.C.: U.S. Government Printing Office, 1931.

Minutes of the Proceedings of the Courts of Inquiry and Court Martial, in Relation to Captain David Porter: Convened at Washington, D.C. On Thursday, the Seventh Day of July, A.D. 1825. Printed by Authority from the Official Record. Washington, D.C.: Davis and Force, 1825.

Murrell, William Meacham. *Cruise of the Frigate Columbia around the World, under the Command of Commodore George C. Read, in 1838, 1839, and 1840*. Boston: B. B. Mussey, 1840.

Parker, James L. *Journal of a Cruise Around the World*. Microfilm. Annapolis, Md.: Nimitz Library Archive, U.S. Naval Academy.

Preble, George Henry, ed. *The First Cruise of the United States Frigate Essex, with a Short Account of Her Origin, and Subsequent Career*. Salem, Mass.: Essex Institute, 1870.

Rutland, Robert Allen, ed. *The Papers of James Madison*. Charlottesville, Va.: University Press of Virginia, 1984.

Sawtelle, Joseph G., ed. *John Paul Jones and the Ranger: Portsmouth, New Hampshire, July 12–November 1, 1777, and the Log of the Ranger, November 1, 1777–May 18, 1778*. Portsmouth, N.H.: Portsmouth Marine Society, 1994.

Seitz, Don Carlos. *Paul Jones, His Exploits in English Seas during 1778–1780, Contemporary Accounts Collected from English Newspapers, with a Complete Bibliography*. New York: E. P. Dutton, 1917.

Stewart, Charles W., ed. *John Paul Jones Commemoration at Annapolis, April 24, 1906*. Washington, D.C.: U.S. Government Printing Office, 1907.

Stewart, C. S. *A Visit to the South Seas, in the U.S. Ship Vincennes, during the Years 1829 and 1830; with Scenes in Brazil, Peru, Manila, the Cape of Good Hope, and St. Helena*. New York: J. P. Haven, 1831.

Stocqueler, J. H., ed. *Memoirs and Correspondence of Major-General Sir William Nott, Volume I*. London: Hurst & Blackett, 1854.

Taylor, Fitch W. *A Voyage Round the World, and Visits to Various Foreign Countries, in the United States Frigate Columbia.* New York: D. Appleton, 1843.

U.S. Congress. *American State Papers: Naval Affairs.* 4 vols. Washington, D.C.: U.S. Government Printing Office, 1832–61.

U.S. Congress. *American State Papers: Foreign Relations.* 6 vols. Washington, D.C.: U.S. Government Printing Office, 1832–59.

U.S. Congress. *The Public Statutes at Large of the United States of America, 1789–1875, vols. 1–3.* Washington, D.C.: U.S. Government Printing Office, 1845–46.

U.S. Office of Naval Records and Library. *Register of Officer Personnel, United States Navy and Marine Corps, and Ships' Data, 1801–1807.* Washington, D.C.: U.S. Government Printing Office, 1945.

Warriner, Francis. *Cruise of the United States Frigate Potomac Round the World, during the Years 1831–34 Embracing the Attack on Quallah-Battoo.* New York: Leavitt, Lord, 1835.

Wharton, Francis, John Bassett Moore, and Jared Sparks, eds. *The Revolutionary Diplomatic Correspondence of the United States.* Washington, D.C.: U.S. Government Printing Office, 1889.

Wingfield, David, Don Bamford, and Paul Carroll. *Four Years on the Great Lakes, 1813–1816: The Journal of Lieutenant David Wingfield, Royal Navy.* Toronto: Natural Heritage Books, 2009.

Newspapers

Connecticut Courant (US)
Daily National Intelligencer (US)
Liverpool Mercury (UK)
London Evening Post (UK)
The Morning Post and Daily Advertiser (US)
National Advocate (US)
Niles Weekly Register (US)
Norfolk Herald (US)
St. James Chronicle/British Evening Post (UK)
The Times (UK)
The War (US)

Secondary Sources

Alden, Carroll Storrs. *Lawrence Kearny, Sailor Diplomat.* Princeton, N.J.: Princeton University Press, 1936.

Aldrich, M. Almy. *History of the United States Marine Corps.* Boston: H. L. Shepard, 1875.

Allen, Gardner Weld. *Our Naval War with France.* Boston: Houghton Mifflin, 1909.

———. *Our Navy and the Barbary Corsairs.* Boston: Houghton Mifflin, 1905.

———. *Our Navy and the West Indian Pirates.* Salem, Mass.: Essex Institute, 1929.

Allen, Joseph. *Battles of the British Navy, Volume 1.* London: Henry G. Bohn, 1852.

Armstrong, Benjamin, ed. *21st Century Sims: Innovation, Education, and Leadership for the Modern Era.* Annapolis, Md.: Naval Institute Press, 2015.

———. "D— All of the Above: Connecting 21st Century Naval Doctrine to Strategy." *Infinity Journal* 4, no. 4 (Summer 2015): 13–17.

———. "Daring Moves on the Niagara." *Naval History* 27, no. 5 (October 2013): 36–42.

———. "An Act of War on the Eve of Revolution." *Naval History* 30, no. 1 (February 2016): 42–48.

Arthur, Brian. *How Britain Won the War of 1812: The Royal Navy's Blockades of the United States, 1812–1815*. Suffolk, UK: Boydell Press, 2011.

Allison, Robert J. *Stephen Decatur: American Naval Hero*. Boston: University of Massachusetts Press, 2005.

Barnett, Roger W. *Navy Strategic Culture: Why the Navy Thinks Differently*. Annapolis, Md.: Naval Institute Press, 2009.

Barrow, Clayton R. *America Spreads Her Sails: U.S. Seapower in the 19th Century*. Annapolis, Md.: Naval Institute Press, 1973.

Bauer, K. Jack. *Surfboats and Horse Marines; U.S. Naval Operations in the Mexican War, 1846–48*. Annapolis, Md.: Naval Institute Press, 1969.

Beach, Edward. "The Court-Martial of Commodore David Porter." U.S. Naval Institute *Proceedings* 23, no. 4 (December 1907): 1392–1402.

Beeler, John. "The State of Naval History." *Historically Speaking* 11, no. 4 (September 2010): 13.

———. "Maritime Policing and the Pax Britannica: The Royal Navy's Anti-Slavery Patrol in the Caribbean 1828–1848." *The Northern Mariner* 26, no. 1 (January 2006): 1–20.

Belohlavek, John M. "'Let the Eagle Soar!': Democratic Constraints on the Foreign Policy of Andrew Jackson." *Presidential Studies Quarterly* 10, no. 1 (1980): 36–50.

Berube, Claude, and John Rodgaard. *A Call to the Sea: Captain Charles Stewart of the USS Constitution*. Washington, D.C.: Potomac Books, 2005.

Biographical Directory of the United States Congress. "John Langdon." http://bioguide .congress.gov/scripts/biodisplay.pl?index=L000067.

"Biographical Memoir of Commodore Dale," *The Port Folio* 3, no. 6 (June 1814): 499–515.

Black, Jeremy. *The War of 1812 in the Age of Napoleon*. Norman: University of Oklahoma Press, 2009.

Boot, Max. *The Savage Wars of Peace: Small Wars and the Rise of American Power*. New York: Basic Books, 2014.

Bowen-Hassell, E. Gordon, Dennis Michael Conrad, and Mark L. Hayes. *Sea Raiders of the American Revolution: The Continental Navy in European Waters*. Washington, D.C.: Naval Historical Center, 2003.

Bradford, James C., ed. *America, Sea Power, and the World*. Chichester, UK: Wiley Blackwell, 2016.

———, ed. *Captains of the Old Steam Navy: Makers of the American Naval Tradition, 1840–1880*. Annapolis, Md.: Naval Institute Press, 1986.

———, ed. *Command under Sail: Makers of the American Naval Tradition, 1775–1850*. Annapolis, Md.: Naval Institute Press, 1985.

———. "John Paul Jones and Guerre de Razzia." *The Northern Mariner* 13, no. 4 (October 2003): 1–15.

Bradlee, Francis Boardman Crowninshield. *Piracy in the West Indies and Its Suppression*. Salem, Mass.: Essex Institute, 1923.

Budiansky, Stephen. *Perilous Fight: America's Intrepid War with Britain on the High Seas, 1812–1815*. New York: Alfred A. Knopf, 2010.

Buker, George. *Swamp Sailors: Riverine Warfare in the Everglades, 1835–1842*. Gainesville: University of Florida Press, 1975.

Burgess, Robert Forrest. *Ships Beneath the Sea: A History of Subs and Submersibles*. New York: McGraw-Hill, 1975.

Butler, Stuart Lee. *Defending the Old Dominion: Virginia and Its Militia in the War of 1812*. Lanham, Md.: University Press of America, 2013.

Cable, James. *Gunboat Diplomacy 1919–1991: Political Applications of Limited Naval Force*. New York: St. Martin's Press, 1994.

Calderhead, William. "Naval Innovation in Crisis: War in the Chesapeake, 1813." *American Neptune*, no. 36 (1976): 206–21.

Callo, Joseph F. *John Paul Jones: America's First Sea Warrior*. Annapolis, Md.: Naval Institute Press, 2006.

Calvert, James F. *The Naval Profession*. New York: McGraw-Hill, 1971.

Campbell, John F. "Pepper, Pirates, and Grapeshot." *American Neptune*, no. 21 (October 1961): 292–301.

Chapelle, Howard Irving. "Fulton's 'Steam Battery': Blockship and Catamaran." *Contributions from the Museum of History and Technology: Paper 39*. Washington, D.C.: U.S. Government Printing Office, 1964.

———. *The History of American Sailing Ships*. New York: W. W. Norton, 1935.

———. *The History of the American Sailing Navy: The Ships and Their Development*. New York: W.W. Norton, 1949.

Cole, Bernard D. *Gunboats and Marines: The United States Navy in China, 1925–1928*. Newark, Del.: University of Delaware Press, 1983.

Colomb, P. H. *Naval Warfare, Its Ruling Principles and Practice Historically Treated*. Classics of Sea Power Edition. Annapolis, Md.: Naval Institute Press, 1990.

Cooper, James Fenimore. *Lives of Distinguished American Naval Officers*. Philadelphia: Carey and Hart, 1846.

Corbett, Julian Stafford. *Some Principles of Maritime Strategy*. London: Longmans, Green, 1918.

Corum, James S., and Wray R. Johnson. *Airpower in Small Wars: Fighting Insurgents and Terrorists*. Lawrence, Kans.: University Press of Kansas, 2003.

Crawford, M. Mac Dermot. *The Sailor Whom England Feared; Being the Story of Paul Jones, Scotch Naval Adventurer and Admiral in the American and Russian Fleets*. New York: Duffield, 1913.

Crawford, Michael, and Christine Hughes. *The Reestablishment of the Navy, 1787–1801: Historical Overview and Select Bibliography*. Washington, D.C.: Naval Historical Center, 1995.

Cunliffe, Tom. *Pilot Cutters Under Sail: Pilots and Pilotage in Britain and Northern Europe*. Barnsley, UK: Seaforth Publishing, 2013.

Cusick, James G. *The Other War of 1812: The Patriot War and the American Invasion of Spanish East Florida*. Gainesville, Fla.: University Press of Florida, 2003.

Daughan, George C. *1812: The Navy's War*. New York: Basic Books, 2011.

———. *If by Sea: The Forging of the American Navy—from the American Revolution to the War of 1812*. New York: Basic Books, 2008.

Davis, Junius. *Some Facts about John Paul Jones*. Raleigh, N.C.: Presses of Edwards and Broughton, 1906.

DeConde, Alexander. *The Quasi-War: The Politics and Diplomacy of the Undeclared War with France 1797–1801*. New York: Scribner, 1966.

DeKay, James Tertius. *The Battle of Stonington: Torpedoes, Submarines, and Rockets in the War of 1812*. Annapolis, Md.: Naval Institute Press, 1990.

De Koven, Anna. *The Life and Letters of John Paul Jones*. New York: C. Scribner's Sons, 1913.

Denzin, Norman K., and Yvonna S. Lincoln, eds. *The SAGE Handbook of Qualitative Research*. Thousand Oaks, Calif.: Sage Publications, 2005.

Dictionary of American Naval Fighting Ships. Washington, D.C.: Naval History & Heritage Command. http://www.history.navy.mil/research/histories/ship-histories/danfs.html.

Duboi, Laurent. *A Colony of Citizens: Revolution and Slave Emancipation in the French Caribbean, 1787–1804*. Chapel Hill: University of North Carolina Press, 2004.

Dudley, William S., and Michael J. Crawford, eds. *The Early Republic and the Sea: Essays on Naval and Maritime History of the Early United States*. Washington, D.C.: Brassey's, 2001.

Dull, Jonathan R. *American Naval History, 1607–1865: Overcoming the Colonial Legacy*. Lincoln, Nebr.: University of Nebraska Press, 2012.

Dunnavent, R. Blake. *Brown Water Warfare: The U.S. Navy in Riverine Warfare and the Emergence of a Tactical Doctrine, 1775–1970*. Gainesville: University Press of Florida, 2003.

Dupont, S. F. "The War with Mexico: The Cruise of the U.S. Ship Cyane During the Years 1845–1848." U.S. Naval Institute *Proceedings* 8, no. 3 (August 1882): 419–37.

Earle, Peter. *The Pirate Wars*. New York: Thomas Dunne Books, 2003.

Elias, Norbert, with R. Moelker and Stephen Mennell, eds. *The Genesis of the Naval Profession*. Dublin, Ireland: University College Dublin Press, 2007.

Farragut, Loyall. *The Life of David Glasgow Farragut, First Admiral of the United States Navy, Embodying His Journal and Letters*. New York: D. Appleton, 1879.

Field, Cyril. *The Story of the Submarine: From the Earliest Ages to the Present Day*. London: Sampson Low, Marston, 1908.

Folkard, Henry Coleman. *The Sailing Boat: A Description of English and Foreign Boats*. London: Hurst and Son, 1853.

Fowler, William M. *Jack Tars and Commodores: The American Navy, 1783–1815*. Boston: Houghton Mifflin, 1984.

———. *Silas Talbot: Captain of Old Ironsides*. Mystic, Conn.: Mystic Seaport Museum, 1995.

———. "The Navy's Barbary War Crucible." *Naval History* 19, no. 4 (August 2005): 55–59.

Fredriksen, John C. *American Military Leaders: From Colonial Times to the Present*. Santa Barbara, Calif.: ABC-CLIO, 1999.

Frierson, J. Gordon. "The Yellow Fever Vaccine: A History." *The Yale Journal of Biology and Medicine* 83, no. 2 (June 2010): 77–85.

Fulton, Robert. *Torpedo War and Submarine Explosions: The Liberty of the Seas Will Be the Happiness of the Earth*. New York: William Elliot, 1810.

Gallagher, Mary A. Y. "Charting a New Course for the China Trade: The Late Eighteenth-Century American Model." *American Neptune*, no. 57 (1997): 201–5.

Gilje, Paul A. *Free Trade and Sailors' Rights in the War of 1812*. New York: Cambridge University Press, 2013.

Gilpin, William. *Observations on the River Wye, and Several Parts of South Wales, &c. Relative Chiefly to Picturesque Beauty, Made in the Summer of the Year 1770*. London: R. Blamire, 1789.

Girard, Philippe. *Toussaint Louverture: A Revolutionary Life*. New York: Basic Books, 2016.

Goldenberg, Joseph A. "Blue Lights and Infernal Machines: The British Blockade of New London." *The Mariner's Mirror* 61, no. 4 (January 1, 1975): 385–97.

Gough, Barry M. *Fighting Sail on Lake Huron and Georgian Bay: The War of 1812 and Its Aftermath*. Annapolis, Md.: Naval Institute Press, 2002.

———. "The Influence of Sea Power Upon History Revisited: Vice-Admiral P. H. Colomb, RN." *Royal United Services Institute Journal* 135, no. 2 (Summer, 1990): 55–63.

Grant, James. *John Adams: Party of One*. New York: Farrar, Straus and Giroux, 2005.

Grant, Patrick. "The Resurrection of John Paul Jones." *Naval History* 26, no. 1 (February 2012): 52–57.

Greenert, Jonathan W. "Building on a 200-Year Legacy." U.S. Naval Institute *Proceedings* 138, no. 5 (May 2012): 32–33.

Grenier, John. *The First Way of War: American War Making on the Frontier, 1607–1814*. Cambridge, UK: Cambridge University Press, 2005.

Grodzinski, John R. *Defender of Canada: Sir George Prevost and the War of 1812*. Norman: University of Oklahoma Press, 2013.

Guttridge, Leonard F. *Our Country, Right or Wrong: The Life of Stephen Decatur, the U.S. Navy's Most Illustrious Commander*. New York: Forge, 2006.

Hackemer, Kurt. *The U.S. Navy and the Origins of the Military-Industrial Complex, 1847–1883*. Annapolis, Md.: Naval Institute Press, 2001.

Hagan, Kenneth J. *American Gunboat Diplomacy and the Old Navy, 1877–1889*. Westport, Conn.: Greenwood Press, 1973.

———, ed. *In Peace and War: Interpretations of American Naval History, 1775–1978*. Westport, Conn.: Greenwood Press, 1978.

———. *This People's Navy: The Making of American Sea Power*. New York: Free Press, 1991.

Hamilton, Alexander. "Federalist Papers: No. 11." http://avalon.law.yale.edu/18th_century/fed11.asp.

Hanks, Robert. "Commodore Lawrence Kearny, the Diplomatic Seaman." U.S. Naval Institute *Proceedings* 96, no. 11 (November 1970): 70–73.

Hattendorf, John B., ed. *Doing Naval History: Essays toward Improvement*. Newport, R.I.: Naval War College Press, 1995.

———. *Talking about Naval History: A Collection of Essays*. Newport, R.I.: Naval War College Press, 2011.

———. "The United States Navy in the Twenty-First Century: Thoughts on naval theory, strategic constraints, and opportunities." *Mariner's Mirror* 97, no. 1 (February 2011): 285–97.

Haynes, Peter D. *Toward a New Maritime Strategy: American Naval Thinking in the Post-Cold War Era*. Annapolis, Md.: Naval Institute Press, 2015.

Head, David. *Privateers of the Americas: Spanish American Privateering from the United States and the Influence of Geopolitics in the Early Republic*, Athens: University of Georgia Press: 2015.

Healy, David. *Gunboat Diplomacy in the Wilson Era: The U.S. Navy in Haiti, 1915–1816*. Madison: University of Wisconsin Press, 1876.

Hickey, Donald R. *The War of 1812: A Forgotten Conflict*. Urbana: University of Illinois Press, 1989.

———. *Don't Give Up the Ship!: Myths of the War of 1812*. Urbana: University of Illinois Press, 2012.

Hitchens, Christopher. *Arguably: Essays*. New York: Twelve, 2011.

Hodge, Carl Cavanagh, and Cathal J. Nolan. *U.S. Presidents and Foreign Policy from 1789 to the Present*. Santa Barbara, Calif.: ABC-CLIO, 2007.

Howard, Michael. "The Use and Abuse of Military History." *Royal United Services Institute Journal* 107, no. 625 (February 1, 1962): 4–10.

Hunter, Mark C. *Policing the Seas: Anglo-American Relations and the Equatorial Atlantic, 1819–1865*. St. Johns, Newfoundland, Canada: International Maritime Economic History Association, 2008.

Hutcheon, Wallace. *Robert Fulton: Pioneer of Undersea Warfare*. Annapolis, Md.: Naval Institute Press, 1981.

Jampoler, Andrew C. A. *Embassy to the Eastern Courts: America's Secret First Pivot toward Asia, 1832–37*. Annapolis, Md.: Naval Institute Press, 2015.

Johnson, Robert Erwin. *Far China Station: The U.S. Navy in Asian Waters, 1800–1898*. Annapolis, Md.: Naval Institute Press, 1979.

———. *Thence Round Cape Horn: The Story of United States Naval Forces on Pacific Station, 1818–1923*. Annapolis, Md.: Naval Institute Press, 1963.

Karsten, Peter. *The Naval Aristocracy: The Golden Age of Annapolis and the Emergence of Modern American Navalism*. Annapolis, Md.: Naval Institute Press, 2008.

Kilmeade, Brian, and Don Yaeger. *Thomas Jefferson and the Tripoli Pirates: The Forgotten War That Changed American History*. New York: Sentinel, 2015.

Kirsch, Peter. *Fireship: The Terror Weapon of the Age of Sail*. Annapolis, Md.: Naval Institute Press, 2009.

Knox, Dudley Wright. *A History of the United States Navy*. New York: G. P. Putnam's Sons, 1948.

Lambert, Andrew D. "Naval History: Division or Dialogue?" *Historically Speaking* 11, no. 4 (2010): 9–11.

———. *The Challenge: America, Britain and the War of 1812*. London: Faber and Faber, 2012.

———. "The Limits of Naval Power: The Merchant Brig Three Sisters, Riff Pirates, and British Battleships." In *Piracy and Maritime Crime: Historical and Modern Case Studies*, edited by Bruce Elleman, Andrew Forbes, and David Rosenberg, 173–90. Newport, R.I.: Naval War College Press, 2010.

Lambert, Frank. *The Barbary Wars: American Independence in the Atlantic World*. New York: Hill and Wang, 2005.

Laughton, John Knox. *Studies in Naval History: Biographies*. London: Longmans, Green, 1887.

Lee, Sidney, ed. *Dictionary of National Biography, Volume LV*. London: Smith, Elder, 1898.

Leiner, Frederick C. *The End of Barbary Terror: America's 1815 War against the Pirates of North Africa*. Oxford, UK: Oxford University Press, 2006.

———. *Millions for Defense: The Subscription Warships of 1789*. Annapolis, Md.: Naval Institute Press, 2000.

———. "Searching for Nelson's Iconic Quote." *Naval History* 26, no. 4 (August 2012): 48–52.

Long, David F. *Gold Braid and Foreign Relations: Diplomatic Activities of U.S. Naval Officers, 1798–1883*. Annapolis, Md.: Naval Institute Press, 1988.

———. *"Mad Jack": The Biography of Captain John Percival, USN, 1779–1862*. Westport, Conn.: Greenwood Press, 1993.

———. "'Martial Thunder': The First Official American Armed Intervention in Asia." *Pacific Historical Review* 42, no. 2 (1973): 143–62.

———. *Nothing Too Daring: A Biography of Commodore David Porter, 1780–1843*. Annapolis, Md.: Naval Institute Press, 1970.

———. *Sailor-Diplomat: A Biography of Commodore James Biddle, 1783–1848*. Boston: Northeastern University Press, 1983.

Lorenz, Lincoln. *John Paul Jones: Fighter for Freedom and Glory*. Annapolis, Md.: Naval Institute Press, 2014.

Lowe, Corinne. *Quicksilver Bob: A Story of Robert Fulton*. New York: Harcourt, Brace, 1946.

Lynch, John. *The Spanish American Revolutions, 1808–1826*. New York: Norton, 1986.

Macdermott, A. "Notes: Father of the American Navy." *The Mariner's Mirror* 48, no. 1 (January 1962): 71–72.

Mahan, A. T. *The Influence of Sea Power upon History, 1660–1783*. Boston: Little, Brown, 1890.

Malcomson, Robert. *Lords of the Lake: The Naval War on Lake Ontario, 1812–1814*. Annapolis, Md.: Naval Institute Press, 1998.

———. *Warships of the Great Lakes, 1754–1834*. Annapolis, Md.: Naval Institute Press, 2001.

Marder, Arthur Jacob. *From the Dardanelles to Oran: Studies of the Royal Navy in War and Peace, 1915–1940*. London: Oxford University Press, 1974.

Martelle, Scott. *The Admiral and the Ambassador: One Man's Obsessive Search for the Body of John Paul Jones*. Chicago: Chicago Review Press, 2014.

Matthew, H. C. G., and Brian Harrison. *Oxford Dictionary of National Biography*. Vol. 53. Oxford, UK: Oxford University Press, 2004.

Mattis, James N., and Frank Hoffman. "Future Warfare: The Rise of Hybrid Wars." U.S. Naval Institute *Proceedings* 131, no. 11 (November 2005): 18–19.

McCarthy, Matthew. *Privateering, Piracy, and British Policy in Spanish America, 1810–1830*. Suffolk, UK: Boydell, 2013.

McCranie, Kevin D. *Utmost Gallantry: The U.S. and Royal Navies at Sea in the War of 1812*. Annapolis, Md.: Naval Institute Press, 2011.

McGrath, Tim. *Give Me a Fast Ship: The Continental Navy and America's Revolution at Sea*. New York: NAL Caliber, 2014.

McKee, Christopher. *A Gentlemanly and Honorable Profession: The Creation of the U.S. Naval Officer Corps, 1794–1815*. Annapolis, Md.: Naval Institute Press, 1991.

———. *Edward Preble: A Naval Biography, 1761–1807*. Classics of Naval Literature. Annapolis, Md.: Naval Institute Press, 1996.

McMichael, F. Andrew. *Atlantic Loyalties: Americans in Spanish West Florida, 1785–1810*. Athens: University of Georgia Press, 2008.

Meacham, Jon. *American Lion: Andrew Jackson in the White House*. New York: Random House, 2008.

Mecham, J. Lloyd. *A Survey of United States–Latin American Relations*. Boston: Houghton Mifflin, 1965.

Morison, Samuel Eliot. *John Paul Jones, a Sailor's Biography*. Boston: Little, Brown, 1959.

———. *The Maritime History of Massachusetts, 1783–1860*. New York: Houghton Mifflin, 1921.

Nash, Howard P. *The Forgotten Wars: The Role of the U.S. Navy in the Quasi War with France and the Barbary Wars 1798–1805*. South Brunswick, N.J.: A. S. Barnes, 1968.

Neimeyer, Charles Patrick. *War in the Chesapeake: The British Campaigns to Control the Bay, 1813–1814*. Annapolis, Md.: Naval Institute Press, 2015.

Norton, Louis A. *Captains Contentious: The Dysfunctional Sons of the Brine.* Columbia: University of South Carolina Press, 2009.

Palmer, Michael A. *Stoddert's War: Naval Operations during the Quasi-War with France, 1798–1801.* Annapolis, Md.: Naval Institute Press, 2000.

Paret, Peter, Gordon Craig, Felix Gilbert, eds. *Makers of Modern Strategy: From Machiavelli to the Nuclear Age.* Princeton, N.J.: Princeton University Press, 1986.

Park, Steven. *The Burning of His Majesty's Schooner Gaspee.* Yardley, Penn.: Westholme Publishing, 2016.

Paul, James Balfour, ed. *The Scots Peerage.* Vol. 7. Edinburgh: David Douglas, 1910.

Paullin, Charles Oscar. "Early Voyages of American Naval Vessels to the Orient I." U.S. Naval Institute *Proceedings* 36, no. 2 (June 1910): 429–63.

———. "Early Voyages of American Naval Vessels to the Orient IX: The Cruise of Commodore Read." U.S. Naval Institute *Proceedings* 36, no. 4 (December 1910): 1073–99.

———. *The Navy of the American Revolution; Its Administration, Its Policy, and Its Achievements.* New York: Haskell House Publishers, 1971.

Perrett, Bryan. *Gunboat!: Small Ships at War.* London: Cassell, 2000.

Philbrick, Nathaniel. *Sea of Glory: America's Voyage of Discovery: The U.S. Exploring Expedition, 1838–1842.* New York: Viking, 2003.

Philip, Cynthia Owen. *Robert Fulton: A Biography.* New York: F. Watts, 1985.

Phillips, James Duncan. *Pepper and Pirates: Adventures in the Sumatra Pepper Trade of Salem.* Boston: Houghton Mifflin, 1949.

Porter, David Dixon. *Memoir of Commodore David Porter, of the United States Navy.* Albany, N.Y.: J. Munsell, 1875.

Porter, David, with R. D. Madison and Karen Hamon, eds. *Journal of a Cruise.* Annapolis, Md.: Naval Institute Press, 1986.

Potter, E. B., Chester W. Nimitz, and Henry Hitch Adams, eds. *Sea Power: A Naval History.* Englewood Cliffs, N.J: Prentice-Hall, 1960.

Pratt, Fletcher. *Preble's Boys: Commodore Preble and the Birth of American Sea Power.* New York: Sloane, 1950.

Preston, Antony, and John Major. *Send a Gunboat: The Victorian Navy and Supremacy at Sea, 1854–1904.* London: Conway Maritime, 2007.

Putnam, George Granville. *Salem Vessels and Their Voyages: A History of the Pepper Trade with the Island of Sumatra.* Salem, Mass.: Essex Institute, 1924.

Rafuse, Ethan S. "'Little Phil,' a 'Bad Old Man,' and the 'Gray Ghost': Hybrid Warfare and the Fight for the Shenandoah Valley, August–November 1864. *Journal of Military History* 81, no. 3 (July 2017): 775–801.

Reynolds, J. N. *Voyage of the United States Frigate Potomac under the Command of Commodore John Downes, during the Circumnavigation of the Globe, in the Years 1831, 1832, 1833, and 1834.* New York: Harper & Bros., 1835.

Rider, Hope S. *Valour Fore & Aft, Being the Adventures of the Continental Sloop Providence, 1775–1779, Formerly Flagship Katy of Rhode Island's Navy.* Annapolis, Md.: Naval Institute Press, 1977.

Rodríguez O., Jaime E. *The Independence of Spanish America.* Cambridge, UK: Cambridge University Press, 1998.

Roosevelt, Theodore. *The Naval War of 1812, Or, The History of the United States Navy during the Last War with Great Britain: To Which Is Appended an Account of the Battle of New Orleans, Part II.* New York: G. P. Putnam's Sons, 1900.

Sale, Kirkpatrick. *The Fire of His Genius: Robert Fulton and the American Dream*. New York: Free Press, 2001.

Schroeder, John H. *Commodore John Rodgers: Paragon of the Early American Navy*. Gainesville: University Press of Florida, 2006.

———. *Shaping a Maritime Empire: The Commercial and Diplomatic Role of the American Navy, 1829–1861*. Westport, Conn.: Greenwood Press, 1985.

———. *The Battle of Lake Champlain: A "Brilliant and Extraordinary Victory."* Norman: University of Oklahoma Press, 2015.

Schubert, Frank N. *Other Than War: The American Military Experience and Operations in the Post–Cold War Decade*. Washington, D.C.: Joint History Office, Office of the Chairman of the Joint Chiefs of Staff, 2013.

Scott, Winfield. *Memoirs of Lieut.-General Scott, LL.D.* New York: Sheldon, 1861.

Sherburne, John Henry. *Life and Character of the Chevalier John Paul Jones: A Captain in the Navy of the United States, during Their Revolutionary War*. New York: Wilder & Campbell, 1825.

Sherwood, John Darrell. *War in the Shallows: U.S. Navy Coastal and Riverine Warfare in Vietnam, 1965–1968*. Washington, D.C.: Naval History and Heritage Command, 2015.

Shomette, Donald. "Infernal Machines: Submarine and Torpedo Warfare in the War of 1812." *Sea History*, no. 141 (Winter 2012): 18–22.

Skaggs, David Curtis. *Oliver Hazard Perry: Honor, Courage, and Patriotism in the Early U.S. Navy*. Annapolis, Md.: Naval Institute Press, 2006.

———. *Thomas Macdonough: Master of Command in the Early U.S. Navy*. Annapolis, Md.: Naval Institute Press, 2003.

Skaggs, David Curtis, and Gerard T. Altoff. *A Signal Victory: The Lake Erie Campaign, 1812–1813*. Annapolis, Md.: Naval Institute Press, 1997.

Smith, Charles R. *Marines in the Frigate Navy*. Washington, D.C.: History Division, U.S. Marine Corps, 2006.

Smith, Gene A. *For the Purposes of Defense: The Politics of the Jeffersonian Gunboat Program*. Newark: University of Delaware Press, 1995.

Smith, J. A. "A Man with a Country: Sketch of the Life of Rear Admiral Gregory, of the United States Navy." *The Midland Monthly* 10, no. 2 (August 1898): 156–60.

Smith, Richard Norton. *Patriarch: George Washington and the New American Nation*. New York: Houghton Mifflin, 1993.

Soley, J. R., ed. "The Autobiography of Commodore Charles Morris." *U.S. Naval Institute Proceedings* 6, no. 12 (April 1880): 115–219.

Sprout, Harold H., and Margaret T. Sprout. *The Rise of American Naval Power: 1776–1918*. 2nd ed. Princeton, N.J.: Princeton University Press, 1967.

Swartz, Peter M. *U.S.-Greek Naval Relations Begin: Antipiracy Operations in the Aegean Sea*. Alexandria, Va.: Center for Naval Analyses, 2003.

Symonds, Craig L. *Decision at Sea: Five Naval Battles That Shaped American History*. Oxford, UK: Oxford University Press, 2005.

———. *Navalists and Antinavalists: The Naval Policy Debate in the United States, 1785–1827*. Newark: University of Delaware Press, 1980.

———. *The U.S. Navy: A Concise History*. New York: Oxford University Press, 2016.

Thomas, Evan. *John Paul Jones: Sailor, Hero, Father of the American Navy*. New York: Simon & Schuster, 2003.

Thornton, Rod. *Asymmetric Warfare: Threat and Response in the Twenty-First Century*. Cambridge, UK: Polity Press, 2007.

Till, Geoffrey. *Asia's Naval Expansion: An Arms Race in the Making?* Adelphi 432–433. London: The International Institute for Strategic Studies, 2012.

———. *Seapower: A Guide for the Twenty-First Century*. 3rd ed. London: Routledge, 2013.

Toll, Ian W. *Six Frigates: The Epic History of the Founding of the U.S. Navy*. New York: W.W. Norton, 2006.

Tucker, Glenn. *Dawn Like Thunder: The Barbary Wars and the Birth of the U.S. Navy*. Indianapolis, Ind.: Bobbs-Merrill, 1963.

Tucker, Spencer. *The Jeffersonian Gunboat Navy*. Columbia: University of South Carolina Press, 1993.

———. *Stephen Decatur: A Life Most Bold and Daring*. Annapolis, Md.: Naval Institute Press, 2005.

Turner, Wesley B. *British Generals in the War of 1812: High Command in the Canadas*. Montreal: McGill-Queen's University Press, 1999.

Ukman, Jason. "U.S. Sailors Buried in Libya Will Stay There, for Now." https://www.washingtonpost.com/blogs/checkpoint-washington/post/sailors-buried-in-libya-will-stay-in-libya-for-now/2011/11/28/gIQAMxnlDO_blog.html?noredirect=on&utm_term=.9e6316d6543f.

Utt, Ronald D. *Ships of Oak, Guns of Iron: The War of 1812 and the Forging of the American Navy*. Washington, D.C.: Regnery History, 2012.

Welsh, William J., and David C. Skaggs, eds. *War on the Great Lakes: Essays Commemorating the 175th Anniversary of the Battle of Lake Erie*. Kent, Ohio: Kent State University Press, 1991.

Wheelan, Joseph. *Jefferson's War: America's First War on Terror, 1801–1805*. New York: Carroll & Graf, 2003.

Wheeler, Richard. *In Pirate Waters*. New York: Crowell, 1969.

Whitehorne, Joseph. *The Battle for Baltimore, 1814*. Baltimore, Md.: Nautical & Aviation Publishing, 1997.

Whipple, A. B. C. *To the Shores of Tripoli: The Birth of the U.S. Navy and Marines*. New York: Morrow, 1991.

Yanik, Anthony. *The Fall and Recapture of Detroit in the War of 1812: In Defense of William Hull*. Detroit, Mich.: Wayne State University Press, 2001.

Yoshihara, Toshi, and James Holmes. *Red Star Over the Pacific: China's Rise and the Challenge to U.S. Maritime Strategy*. Annapolis, Md.: Naval Institute Press, 2011.

Unpublished Secondary Sources

Fithian, Charles. "'For the Common Defense,' 'Infernals,' and a 'Marauding Species of War': The War of 1812 in Delaware." Unpublished conference paper, Enemies to Allies Conference, Annapolis, Md., 2013.

O'Neill, Patrick L. "Between the Burning and Bombardment: The Potomac Squadron and the Battle of the White House." Unpublished conference paper, Enemies to Allies Conference, Annapolis, Md., 2013.

Slaughter, Joseph P. "A Navy in the New Republic: Strategic Visions of the United States Navy, 1783–1812." Unpublished master's thesis, University of Maryland, 2006.

Tomblin, Barbara B. "From Sail to Steam: The Development of Steam Technology in the United States Navy, 1838–1865." Unpublished PhD dissertation, Rutgers University, 1988.

Index

Page numbers in *italics* indicate illustrations.

Niagara River: Black Rock cutting-out operation, 75–77, 90–91, 92, 195; raiding and cutting-out operations on, 73, 90–92
Nonsuch, 127–28
North Africa, 54. *See also* Barbary Coast
Nyah-Heit, Po, 178, 179
Nymph, 47–48, 49

Oneida Indians, 81–82, 83, 92
Ontario, Lake: balance of power and maritime supremacy on, 77, 81, 88–89, 90–91; Burlington Races, 77–78, 215n24; Kingston blockade, 89; local watermen as volunteers for naval service, 11, 78, 82, 83, 85, 87, 92; naval arms race on, 73, 77–78; Oswego assault, 81; Presqu'ile schooner burning mission, 87–89, 91, 216n68; Quinte Bay operation and capture of Gregory, 89–90; raiding and cutting-out operations on, 11, 73, 90–92; Royal Navy Kingston supply transport disruption, 85–87, 91; Sandy Creek engagement, 81–84, 91, 92, 215n39; shipyards on shore of building gunboats and schoolers, 87–89; torpedo operations on, 115–16, 120
Ottoman Empire, 54

Palmer, James S., 176–77
Palmer, Michael, 42, 43, 210n33
Parke, Matthew, 18, 22, 23–24, 35
partnerships, relationships, and cooperation: Barbary War operations, 58–59; Black Rock cutting-out operation, 75, 77, 92; importance of to irregular warfare operations, 5–6, 35, 52, 192, 194–95; local allies for local knowledge and intelligence gathering, 34, 35, 52, 78, 82, 83, 85, 87, 88, 92, 101, 192; merchants partnerships in counterpiracy operations, 194; Oneida Indians and Sandy Creek operation, 81–82, 83, 92; Preble efforts during Barbary War, 70–71; Quasi-War operations, 52; relevance of study

of, 13; Royal Navy coordination with West Indies Squadron for counterpiracy operations, 12, 136, 138, 142–45, 148–49, 195; Sandford assistance with cutting-out operation, 48–50, 70; Spanish coordination with West Indies Squadron for counterpiracy operations, 12, 141, 145, 148–49; Sumatra intelligence from local captains, 175, 180; Sumatra local knowledge and intelligence from local people, 169–70, 176, 178, 179–80, 188; torpedo operations and civilian partners, 116–17, 119–21
Patience, 31
Paulding, James, 175, 183, 184, 231n74
Paul Jones, John. *See* Jones, John Paul
peacetime maritime security operations. *See* maritime security operations/ peacetime maritime security operations
Peacock, 129, 130–31, 138
Penny, Joshua, 111–12, 114
Pensacola naval base, 146
pepper, spices, and spice trade: American pepper merchants dominance of Sumatran trade, 170–71, 187; Downes mission and deterrence of further attacks on ships, 170–71; Kuala Batu pepper trade improvement after burning of Muckie, 184; pirate attacks on merchant ships, 12, 172–73, 187; Read diplomatic success and deterrence of further attacks on ships, 189–90; Salem pepper merchants, 150, 170–71, 172; *Sumatra* contract for pepper cargo with Po Quallah, 186. See also *Eclipse*; *Friendship*
Percival, John, 73
Perry, Oliver Hazard, 77, 84, 127–28, 147
Philadelphia: coded message with intelligence on how to retake, 57, 61; Decatur, Sr., command of, 58; Gibraltar operations against Tripoli, 55; grounding and capture of, 56–57, 59, 70, 71; *Intrepid* operation to destroy, 60–63, 94, 194; negotiations